Advances in the Theory and Technology of Physical Separation

Advances in the Theory and Technology of Physical Separation

Guest Editors

Dongfang Lu
Mehdi Safari

MDPI

Basel • Beijing • Wuhan • Barcelona • Belgrade • Novi Sad • Cluj • Manchester

Guest Editors
Dongfang Lu
School of Minerals Processing and Bioengineering
Central South University
Changsha
China

Mehdi Safari
Minerals Processing Division
Mintek
South Africa

Editorial Office
MDPI AG
Grosspeteranlage 5
4052 Basel, Switzerland

This is a reprint of the Special Issue, published open access by the journal *Minerals* (ISSN 2075-163X), freely accessible at: https://www.mdpi.com/journal/minerals/special_issues/598PQ9XIE6.

For citation purposes, cite each article independently as indicated on the article page online and as indicated below:

Lastname, A.A.; Lastname, B.B. Article Title. *Journal Name* **Year**, *Volume Number*, Page Range.

ISBN 978-3-7258-7276-3 (Hbk)
ISBN 978-3-7258-7277-0 (PDF)
https://doi.org/10.3390/books978-3-7258-7277-0

Contents

About the Editors

Dongfang Lu

Dongfang Lu is a Professor and PhD supervisor at Central South University, where he earned his PhD in Mineral Processing Engineering and completed his postdoctoral research in Metallurgical Engineering. His research focuses on efficient material separation, integrating physical, chemical, mathematical, and computational approaches. His work covers enhanced recovery of complex refractory ores; pretreatment and pre-concentration for low-grade refractory resources; computational modeling of fluid and particle dynamics; and development of pretreatment and pre-concentration equipment for low-grade ores based on compound force fields and fluidization theory. He has led national and provincial research projects, including those funded by the NSFC, China Postdoctoral Science Foundation, and Hunan Provincial programs. He has also directed over 50 industry technology projects and participated in an international innovation project. He has published 60 SCI papers in international journals and serves as a reviewer for several SCI journals. He has also authored one monograph, holds six invention patents (as first/second inventor), and received a provincial/ministerial science award. He serves as a review expert for the NSFC, the Ministry of Education Degree Center, and provincial committees.

Mehdi Safari

Mehdi Safari is affiliated with Mintek and has extensive experience in mineral processing, extractive metallurgy, and hydrometallurgical flowsheet development, with a particular focus on flotation, fine particle processing, and sustainable metal recovery. His research and professional interests centre on the beneficiation of low-grade and complex ores, the recovery of critical and precious metals, and the integration of physical separation with hydrometallurgical and adsorption-based technologies. He has been actively involved in the development, evaluation, and scale-up of innovative processing routes for commodities including platinum group metals, base metals, rare earth elements, and iron ores, with a strong emphasis on process optimisation, hydrodynamics, and environmental performance. His work has contributed to several industrial projects, pilot-scale studies, and collaborative research initiatives aimed at improving resource efficiency and reducing the environmental footprint of mineral processing operations. He has published and reviewed scholarly articles in peer-reviewed journals and has served as a supervisor and examiner for postgraduate research in mineral processing and metallurgical engineering. Through his engagement with the industry, research institutions, and international technical conferences, he continues to contribute to advancing applied research and knowledge transfer in sustainable extractive metallurgy.

minerals

itorial

ditorial for the Special Issue "Advances in the Theory and echnology of Physical Separation"

ongfang Lu [1,*] and Mehdi Safari [2,3,*]

1 School of Minerals Processing and Bioengineering, Central South University, Changsha 410083, China
2 Minerals Processing Division, Mintek, Randburg 2125, South Africa
3 Faculty of Engineering and the Built Environment, University of the Witwatersrand, 1 Jan Smuts Ave., Johannesburg 2000, South Africa
* Correspondence: ludongfang@csu.edu.cn (D.L.); mehdis@mintek.co.za (M.S.)

With the continuous consumption of mineral resources, ore grades worldwide are declining, and their increasingly fine-grained, low-grade, mineralogically complex, and highly heterogeneous characteristics have led to a marked rise in the energy consumption and processing costs of mineral separation [1,2]. In this context, physical separation, with its advantages of low operating cost, high throughput, simple flowsheets, and independence from chemical reagents, has become a key pre-concentration approach for reducing the amount of material entering downstream operations, upgrading feed grade, and improving overall resource utilization efficiency. Physical separation techniques not only effectively remove large proportions of gangue in the early stages of processing, thereby reducing the load on subsequent operations, but also play important roles in tailings reduction, solid-waste valorization, and environmentally sustainable mineral processing [3]. With the continuous upgrading and refinement of separation equipment, physical techniques such as grinding classification [4,5], gravity separation [6–8], magnetic separation [9–11], sensor-based sorting [12,13], solid–liquid separation [14,15], and using fluidized bed [16–18] have been increasingly applied throughout mineral processing flowsheets. Their performance in terms of processing capacity, separation accuracy, operational stability, and adaptability has steadily improved, and the range of available equipment has expanded. These technologies enable efficient separation based on differences in density, magnetic susceptibility, dielectric properties, and surface characteristics. Although chemical methods play an important role in modern mineral beneficiation, physical separation techniques such as gravity and magnetic separation remain indispensable complements when flotation performance is limited or when its applicability is constrained [19,20].

This Special Issue of *Minerals*, "Advances in the Theory and Technology of Physical Separation", presents nine contributions that address several key areas of physical separation. They cover flow-field regulation and structural optimization of gravity separation equipment, the design optimization of magnetic separators and magnetic matrices, the influence of grinding and classification parameters on subsequent separation performance, as well as the purification and high-quality preparation of industrial minerals. These studies, spanning equipment innovation, process intensification, and mechanistic understanding, collectively demonstrate the latest advances in physical separation technologies for complex mineral systems, providing important theoretical support and practical guidance for improving resource utilization efficiency and promoting the development of green mining.

1. Advances in Fundamental Understanding

Four papers in this Special Issue, from the fundamental perspectives of fluid dynamics, particle migration, interfacial behavior, and thermal treatment mechanisms, elucidate the key governing processes in physical separation.

Gao et al. (Contribution 1) explored the influence of cross-sectional geometry on the internal flow structure and particle separation performance of spiral concentrators. Via a computational fluid dynamics (CFD) approach, they systematically analyzed how the cross-sectional profile of the mixed-particle zone affected the secondary flow structure and separation behavior of 89.5 μm hematite and quartz particles, and they further optimized line segments through response surface methodology (RSM). The results showed that both the peak radial flux and the average radial velocity of the secondary flow increased with the slope of the corresponding line segments. Experimental validation demonstrated that employing the optimized cross-sectional profile improved the separation performance by nearly 5% when processing −100 + 75 μm hematite–quartz mixed feed. This work reveals the intrinsic connection between cross-sectional geometry, flow-field evolution, and particle migration, providing an important basis for the precise design of spiral concentrator cross-sections and for achieving efficient mineral separation.

Gao et al. (Contribution 2) investigated how wall roughness in the middle zone of a spiral concentrator influences the evolution of the flow field of the slurry and the associated particle separation behavior. By combining surface roughness measurement and numerical experimental methods, they analyzed the response of slurry film thickness, secondary-flow structure, and radial flux. The results show that increasing wall roughness suppresses the intensity of secondary flow, weakens the inward migration of particles toward the inner trough region, and shifts the enrichment zones outward. This study further demonstrates that mid-section wall roughness and feed particle size jointly determine the internal hydrodynamic characteristics and particle stratification in spiral concentrators, thereby providing a theoretical basis for trough material selection, surface-structure optimization, and roughness control during operation.

Guo et al. (Contribution 3) systematically examined how grinding parameters, including grinding concentration, media proportion, and filling ratio, influence the particle-size distribution, interfacial properties, and agglomeration behavior of galena. By integrating laboratory experiments with discrete element method (DEM) simulations, the study clarified how different grinding conditions regulate particle-size composition and showed that appropriate adjustments can significantly change the yields of the +25 μm and −10 μm fractions. Flotation tests demonstrated that excessive generation of −10 μm particles greatly reduces galena recovery because these ultrafine particles exhibit lower contact angles and smaller agglomerate sizes, which weaken their floatability. DEM simulations further indicated that moderately reducing grinding intensity can effectively limit the formation of −10 μm particles. The study establishes a mechanistic pathway in which grinding-induced abrasion intensity governs the generation of ultrafine particles, alters their interfacial behavior, and ultimately affects separation efficiency. This work provides new mechanistic insight into the longstanding challenge of poor physical separation performance for fine minerals and offers practical guidance for the coordinated optimization of grinding and separation processes in industrial applications.

Huang and Zhang (Contribution 4) focused on the development and underlying mechanisms of a combined process for producing high-purity quartz through impurity removal via pretreatment and subsequent metallurgical purification, with particular emphasis on thermal treatment mechanisms. Using river-sand–derived quartz with an SiO_2 content of 92.31% as the raw material, the authors proposed a multi-step flowsheet consisting of magnetic separation, reverse flotation, microwave roasting, and acid leaching, enabling

https://doi.org/10.3390/min16010059

the efficient upgrading and high-value utilization of low-grade quartz. Magnetic separation removed 87.24% of iron-bearing impurities, while flotation, microwave roasting, and leaching further reduced total impurity levels to below 100 ppm. Through SEM, TG–MS, and in situ XRD analyses, the study elucidated the key mechanisms of microwave-assisted impurity removal, including the rupture of fluid inclusions and lattice substitution. The results verify the effectiveness and feasibility of microwave thermal treatment for purifying low-grade quartz. In addition, the optimized magnetic separation and roasting parameters provide useful guidance for the industrial production of high-quality quartz.

Individually, these studies provide new scientific insights into how equipment structural parameters, grinding behavior, and impurity occurrence states influence separation performance.

2. Innovative Techniques and Applications

Five contributions in this Special Issue present technological advances and applied solutions that expand the capabilities and industrial relevance of physical separation processes.

Liu et al. (Contribution 5) demonstrated that a laboratory-scale reflux classifier (RC) could effectively preconcentrate ultrafine ilmenite tailings with a particle size D_{50} of 13.8 μm and TiO_2 grade of 11.62%; a TiO_2 concentrate with a grade of 20.3% and a recovery rate of 82.8% could be obtained, and the separation efficiency reached 50%. The study elucidated the synergistic control mechanisms of fluidized-bed density, hindered settling, and inclined-surface flow on fine particle stratification. Comparative tests showed that magnetic separation performed poorly due to entrainment of magnetic gangue, whereas the RC achieved cleaner separation based on density disparity, significantly improving separation precision and economic efficiency. This approach provides a new, low-cost pathway for the preconcentration of ultrafine particles.

Liu et al. (Contribution 6) addressed the limited processing capacity of transverse-field high-gradient magnetic separators (HGMS) by systematically optimizing the structural parameters of the matrix box, including the diameter of the rod matrix, the filling ratio, the depth of the matrix in the direction of slurry flow. The study showed that these structural factors jointly determined the distribution of the magnetic capture zone, particle capture probability, and the ore unloading efficiency. Finally, an optimal double-layer matrix ring was applied to an industrial transverse-field HGMS, and its inner and outer rings were equipped with matrix boxes with φ3 mm and φ2 mm rods, respectively, which significantly enhanced ilmenite preconcentration efficiency and processing capacity, yielding higher TiO_2 grades and recovery than conventional longitudinal-field HGMS. This work highlights the critical role of structural optimization in improving the magnetic separation performance and provides a clear, engineering-feasible pathway for industrial matrix design.

Shi et al. (Contribution 7) proposed a 1 mm dry flip-flow screening technique to improve the screening of fine ilmenite particles, where conventional systems often suffer from clogged holes and reduced efficiency. Using a laser displacement measurement system, the researchers analyzed the dynamic characteristics of the flip-flow screen and evaluated its performance under different feed-size distributions and different ratios of obstructive particles. The results showed that the 1 mm screening efficiency reached 85.41%, demonstrating a significant improvement in fine-particle classification. The study also developed a pre-throwing tailings process based on 1 mm multi-stage screening, which does not use water and is particularly suitable for dry, water-scarce, or high-altitude permafrost regions. This provides strong potential for practical application.

Xu et al. (Contribution 8) evaluated the effects of smooth and humped flow channels on the recovery of industrial magnetic seeds in a drum magnetic separator. The results showed that under a range of operating conditions, the humped channel consistently achieved

https://doi.org/10.3390/min16010059

higher recovery than the smooth channel, with improvements of up to 3%. Simulations of the magnetic field, flow field, and particle trajectories revealed that the humped channel shifted impurity magnetic seeds from regions of weak magnetic force into those of stronger magnetic force, thereby significantly enhancing their capture efficiency. This work provides a clear structural optimization strategy for drum magnetic separators and offers practical value for improving magnetic flocculation performance in water-treatment applications.

Shafiee et al. (Contribution 9) developed an optimized and scalable synthesis route for high-purity barium sulfate using D-optimal methodology. By systematically varying $BaCl_2$ and sulfate concentrations, feed rate, and reaction temperature, the researchers established and validated the key parameters controlling product yield and density. Under the optimized conditions, the process produced barium sulfate with a density of 4.25 g/cm^3 and an average purity of 96.08%. The method reduces reagent consumption, minimizes secondary waste, and demonstrates strong potential for sustainable and cost-effective scale-up, offering a practical approach for the green preparation of high-purity inorganic salts.

Collectively, these studies show how innovations in device structures, screening strategies, magnetic configurations, and process optimization can translate physical-separation principles into practical, scalable, and industry-ready solutions.

3. Future Directions

The contributions in this Special Issue reflect two major trends in physical separation research: (1) the continuous deepening of fundamental mechanistic investigations into multiphysics processes, such as flow-field evolution, particle migration, interfacial interactions, and impurity occurrence; (2) the effective translation of these mechanistic insights into engineering practice, driving structural optimization and performance enhancement of physical separation equipment, as well as the emergence of scalable and industrially applicable mineral processing technologies. Together, the presented studies demonstrate that physical separation, as a critical pillar of mineral processing, is continuously evolving. It not only retains its irreplaceable role in the beneficiation of traditional ores but also exhibits increasing strategic importance in the treatment of ultrafine and low-grade minerals, the efficient recovery of critical minerals, the revalorization of tailings, and environmentally responsible resource development.

Looking forward, the deep integration of computational multiphysics modeling, intelligent optimization algorithms, high-efficiency physical separation equipment, and green, low-carbon mineral processing flowsheets will serve as a central driving force for the continued advancement of physical separation and for empowering the broader mineral processing value chain. Meanwhile, the convergence of physical separation, multiscale multiphysics simulation, particle–fluid coupling models, intelligent sensing and process digitalization, data-driven process optimization, and environmentally sustainable processing technologies will represent a key direction for future development, guiding the industry toward higher efficiency, lower energy consumption, greater intelligence, and long-term sustainability.

We thank all the authors for their valuable contributions, the reviewers for their thoughtful evaluations, and the editorial team of *Minerals* for their support. We anticipate that the presented studies will serve as valuable references for researchers and practitioners, while also inspiring new directions in the theory and technology of physical separation.

Author Contributions: Writing—original draft preparation, D.L. and M.S.; writing—review and editing, D.L. and M.S. All authors have read and agreed to the published version of the manuscript.

Conflicts of Interest: The authors declare no conflicts of interest.

https://doi.org/10.3390/min1601005

List of Contributions:

1. Gao, S.; Wang, Q.; Zhou, X.; Liu, C.; Shen, Y.; Cui, B. Effect of the Cross-Sectional Geometry of the Mixed Particle Zone on the Spiral Separation Process and Its Structural Optimization. *Minerals* **2024**, *14*, 1251. https://doi.org/10.3390/min14121251.
2. Gao, S.; Zhou, X.; Li, B.; Wang, Q.; Liu, C. Effect of Wall Roughness in the Middle Zone of Spiral Concentrator on the Flow Field Evolution of Hematite–Quartz Slurry and Particle Separation Behaviour. *Minerals* **2025**, *15*, 208. https://doi.org/10.3390/min15030208.
3. Guo, M.; Yang, Y.; Yu, S.; Wu, Y.; Gu, G.; Wang, Y.; Li, Q.; Chen, J. Effects of Grinding Parameters on Galena Particle Size Distribution and Flotation Performance. *Minerals* **2025**, *15*, 618. https://doi.org/10.3390/min15060618.
4. Huang, H.; Zhang, N. Study on the Preparation of High-Quality Quartz and Its Mechanism by Combining Pretreatment with Metallurgy. *Minerals* **2024**, *14*, 1229. https://doi.org/10.3390/min14121229.
5. Liu, Z.; Su, Z.; Liu, B.; Wang, Y.; Zhang, Y.; Zhong, X.; Chen, K.; Hu, X.; Lu, D. Preconcentrating Ultrafine Ilmenite Tailings Using a Laboratory-Scale Reflux Classifier. *Minerals* **2024**, *14*, 1125. https://doi.org/10.3390/min14111125.
6. Liu, J.; Dai, H.; Yu, L.; Wang, C.; Feng, J.; Li, P.; Xu, S. Optimization of the Matrix in a Transverse-Field High-Gradient Magnetic Separator for an Improved Ilmenite Separation. *Minerals* **2025**, *15*, 114. https://doi.org/10.3390/min15020114.
7. Shi, W.; Wang, W.; Mao, P.; Hou, X.; Zhang, S.; Duan, C. Research on the Dry Deep Flip-Flow Screening of Ilmenite and Its Pre-Throwing Tail Processing Technology. *Minerals* **2025**, *15*, 308. https://doi.org/10.3390/min15030308.
8. Xu, S.; Han, H.; Liu, J.; Sun, W.; Qiu, J. Humped Flow Channel in Drum Magnetic Separator Leads to Enhanced Recovery of Magnetic Seeds in Magnetic Flocculation Process. *Minerals* **2025**, *15*, 732. https://doi.org/10.3390/min15070732.
9. Shafiee, A.; Sadatipour, M.; Hoseinian, F.S.; Rezai, B.; Safari, M. The Optimized Synthesis of Barium Sulfate: A Scalable and Sustainable Laboratory Approach Using D-Optimal Design. *Minerals* **2025**, *15*, 621. https://doi.org/10.3390/min15060621.

References

1. Luukkanen, S.; Tanhua, A.; Zhang, Z.; Mollehuara Canales, R.; Auranen, I. Towards waterless operations from mine to mill. *Miner. Eng.* **2022**, *187*, 107793. [CrossRef]
2. Wang, C.; Harbottle, D.; Liu, Q.; Xu, Z. Current state of fine mineral tailings treatment: A critical review on theory and practice. *Miner. Eng.* **2014**, *58*, 113–131. [CrossRef]
3. Poloko, N. Physical separation methods, part 1: A review. *IOP Conf. Ser. Mater. Sci. Eng.* **2019**, *641*, 012023. [CrossRef]
4. Mu, Y.; Shelley, P. The Mill Family Model: A decision-support framework for tailored grinding media selection in semi-autogenous grinding and overflow ball mills. *Results Eng.* **2025**, *28*, 108196. [CrossRef]
5. Fang, X.; Wu, C.; Liao, N.; Zhong, J.; Liu, R.; Wu, J.; Ling, L.; Duan, X. Discrete element analysis of the influence of grinding media shape on the grinding process. *Miner. Eng.* **2025**, *225*, 109222. [CrossRef]
6. Sekhon, S.S.; Sehgal, S. Computational modeling of particle laden flows and optimization techniques for designing spiral separator in mineral processing. *Sep. Purif. Technol.* **2025**, *38*, 136059. [CrossRef]
7. Vakamalla, T.R.; Mangadoddy, N. Rheology-based CFD modeling of magnetite medium segregation in a dense medium cyclone. *Powder Technol.* **2015**, *277*, 275–286. [CrossRef]
8. Lu, D.F.; Zhang, Y.X.; Liu, Z.Q.; Zheng, X.Y.; Wang, Y.H.; Liu, Y.F. A Novel Pulsation Reflux Classifier Used for Enhanced Preconcentration Efficiency of Antimony Oxide Ore. *Minerals* **2025**, *15*, 605. [CrossRef]
9. Hu, Z.; Lu, D.; Zheng, X.; Wang, Y.; Xue, Z.; Xu, S. Development of a high-gradient magnetic separator for enhancing selective separation: A review. *Powder Technol.* **2023**, *421*, 118435. [CrossRef]
10. Ge, W.; Encinas, A.; Araujo, E.; Song, S. Magnetic matrices used in high gradient magnetic separation (HGMS): A review. *Results Phys.* **2017**, *7*, 4278–4286. [CrossRef]
11. Iranmanesh, M.; Hulliger, J. Magnetic separation: Its application in mining, waste purification, medicine, biochemistry and chemistry. *Chem. Soc. Rev.* **2017**, *46*, 5925–5934. [CrossRef] [PubMed]
12. Duan, B.; Bobicki, E.R.; Hum, S.V. Application of microwave imaging in sensor-based ore sorting. *Miner. Eng.* **2023**, *202*, 108303. [CrossRef]

https://doi.org/10.3390/min16010059

13. Yin, J.; Zhu, J.; Zhu, H.; Pan, G.; Zhu, W.; Zeng, Q.; Shi, Q. Intelligent photoelectric identification of coal and gangue—A revie *Measurement* **2024**, *233*, 114723. [CrossRef]
14. Batalović, V. Centrifugal separator, the new technical solution, application in mineral processing. *Int. J. Miner. Process.* **2011**, 1 86–95. [CrossRef]
15. Baust, H.K.; Lam, Z.; Hay, M.; Nirschl, H.; Gleiß, M. Development of a dynamic process model for the mechanical fluid separatic in disk stack centrifuges. *Sep. Purif. Technol.* **2025**, *377*, 134230. [CrossRef]
16. Taguta, J.; Safari, M.; Govender, V.; Chetty, D. Investigating the Amenability of a PGM-Bearing Ore to Coarse Particle Flotatic *Minerals* **2023**, *13*, 698. [CrossRef]
17. Ghorbani, Y.; Zhang, S.E.; Nwaila, G.T.; Bourdeau, J.E.; Safari, M.; Hoseinie, S.H.; Nwaila, P.; Ruuska, J. Dry laboratories Mapping the required instrumentation and infrastructure for online monitoring, analysis, and characterization in the miner industry. *Miner. Eng.* **2023**, *191*, 107971. [CrossRef]
18. Carelse, C.; Manuel, M.; Chetty, D.; Taguta, J.; Safari, M.; Youlton, K. The flotation behaviour of liberated platinum Grou minerals in Platreef ore under reduced reagent conditions. *Miner. Eng.* **2022**, *190*, 107913. [CrossRef]
19. Nzeh, N.S.; Popoola, P.; Okanigbe, D.; Adeosun, S.; Adeleke, A. Physical beneficiation of heavy minerals—Part 1: A state of th art literature review on gravity concentration techniques. *Heliyon* **2023**, *9*, e18919. [CrossRef] [PubMed]
20. Nzeh, N.S.; Popoola, P.A. Physical beneficiation of heavy minerals—Part 2: A state of the art literature review on magnetic an electrostatic concentration techniques. *Heliyon* **2024**, *10*, e32201. [CrossRef] [PubMed]

https://doi.org/10.3390/min1601005

minerals

MDPI

ticle

Humped Flow Channel in Drum Magnetic Separator Leads to Enhanced Recovery of Magnetic Seeds in Magnetic Flocculation Process

aohua Xu [1,2], Haisheng Han [1], Jianguo Liu [2], Wei Sun [1,*] and Jianwei Qiu [3]

1 School of Minerals Processing and Bioengineering, Central South University, Changsha 410083, China; 13172011895@163.com (S.X.); hanhai5086@csu.edu.cn (H.H.)
2 Institute of Resources Utilization and Rare Earth Development, Guangdong Academy of Sciences, Guangzhou 510650, China; liujianguo@grre.gd.cn
3 School of Environmental Science and Engineering, Processing and Bioengineering, Guangdong University of Technology, Guangzhou 510006, China; 18127391473@163.com
* Correspondence: sunmenghu@mail.csu.edu.cn

Abstract: This study examines the effect of smooth and humped flow channels on the recovery of industrial magnetic seeds in a drum magnetic separator. The results demonstrate that under varying feeding slurry quantities and drum rotational speeds, the humped channel consistently achieves higher recovery rates compared with the smooth channel, with an improvement of up to 3%. Scanning electron microscopy and vibrating sample magnetometry analyses of the samples reveal the presence of a small amount of impurities (predominantly consisting of elements, such as Al, Si, and Ti) in the industrial magnetite magnetic particles. These impurities exhibit lower magnetization, leading to reduced capture efficiency in the conventional smooth-channel drum magnetic separator. Simulations of the magnetic field, flow field, and particle trajectory indicate that the magnetic field force at the bottom of the smooth channel is only 0.6 $kg^2/(m \cdot s^4 \cdot A^2)$, i.e., approximately 18 times lower than that at the roller surface. The incorporation of a humped channel shifts the impure magnetic seeds from a region with low magnetic field force to a region with higher magnetic field force, significantly enhancing the capture efficiency of the impure magnetic seeds.

Keywords: drum magnetic separator; humped flow channel; magnetic seeds; magnetite; impurity

1. Introduction

Magnetic flocculation wastewater treatment technology can be used as a novel approach to water treatment that combines the advantages of traditional flocculation technology and magnetic separation technology [1,2]. It has gained widespread attention and application in recent years [3]. The principle of this technology is the addition of magnetic seeds at its core, which, under the network-trapping action of flocculants, aggregate with suspended pollutants in the wastewater to form magnetic flocs. These magnetic flocs are then separated from the water through gravity settling or the application of an external magnetic field by utilizing the high density or high magnetic properties of the magnetic seeds [4,5]. This process achieves the purification of water and the removal of pollutants. This technology effectively accelerates the settling speed of flocs, enhances the removal efficiency of solid particles in wastewater, and addresses issues that are commonly encountered in traditional sedimentation methods, such as low surface load, slow settling speed, large land occupation, and significant water quality fluctuations, through rapid separation via an external magnetic field. Furthermore, magnetic flocculation technology exhibits good selectivity and adjustability, enabling the selection of different flocculants based on the characteristics of various water qualities and pollutant types and thus optimizing the treatment effect [6]. This technology has demonstrated superior performance in removing heavy metal ions [7], organic pollutants [8], oil [9], bacteria [10], and other substances [11] from water, making it particularly suitable for the treatment of difficult-to-handle or cost-prohibitive industrial wastewater and municipal sewage.

To reduce treatment costs and enhance the environmental protection of the system in this process, magnetic seeds must be efficiently recovered and reused using magnetic separation equipment. Therefore, improving the structure of magnetic separation equipment and enhancing the recovery of magnetic seeds are of considerable importance for the advancement of this technology. The literature has reported on magnetic seed recovery equipment, which includes electromagnetic high-gradient [12], disc-permanent [13], and drum-permanent magnetic separators [14]. The first is suitable for recovering fine and weakly magnetic seeds (e.g., modified magnetic seeds, such as limonite, hematite, and ilmenite), while the latter two are suitable for recovering coarser and stronger magnetic seeds (e.g., magnetite, steel slag, and pure iron powder). Tang et al. [15] investigated the influence of separation chamber shape in dry magnetic separators, demonstrating that a wavy-shaped separation chamber (WSC) promotes airflow turbulence and vortex generation. This design effectively disperses aggregated fine particles, improving the recovery of magnetite and high-grade intergrowths while enhancing the overall separation efficiency. Their work highlights the significant role of chamber shape in manipulating particle behavior and separation outcomes. Wang et al. [16] focused on optimizing magnetic field gradients in a wet low-intensity magnetic separator (LIMS), proposing an innovative precise low-intensity magnetic separator (PLIMS) with a gradient magnet assembly (GMA). The GMA is designed to gradually increase the remnant flux density along the slurry flow direction, enabling "precise magnetic force" application to magnetic particles. This approach balances the capture of high-liberation magnetite and low-grade intergrowths, outperforming uniform magnet assemblies (UMAs) in separation performance. Their work emphasizes the potential of tailored magnetic field gradients to enhance both concentrate grade and recovery efficiency. At present, the magnetic seeds that are widely used in industrial wastewater and municipal sewage treatment are strongly magnetic magnetite, and the primary recovery equipment is the drum magnetic separator [4]. Magnetite exhibits advantages, such as its abundant sources, simple preparation process, low production cost, and high recycling rate, compared with other magnetic seeds. However, a relatively high loss rate of magnetic seeds is frequently observed when using traditional drum magnetic separators for mineral processing to recover magnetic seeds [17], inadvertently increasing the cost of wastewater treatment and possibly leading to secondary pollution of the water.

The current study focuses on the issue of magnetic seed loss that is commonly encountered in traditional drum magnetic separators through conducting research on optimizing the flow channel structure. First, industrial condition experiments were performed to compare the effects of humped and traditional smooth channels on magnetic seed recovery. Subsequently, scanning electron microscopy (SEM) and vibrating sample magnetometry (VSM) were conducted to analyze the products of industrial magnetite magnetic seed separation and determine the primary causes of high seed loss rates. Finally, magnetic field and flow field simulations, particle body force calculations, and finite-/discrete-element coupled particle trajectory simulations were utilized to reveal the influence mechanism of the shape of a flow channel on magnetic seed recovery. This research will provide theoretical and practical guidance for enhancing the recovery of magnetic seeds in the magnetic flocculation process.

2. Materials, Equipment, and Methods

2.1. Raw Magnetic Seeds

The raw magnetic seeds used in this study originated from an industrial magnetite sample used in a wastewater treatment company in Zhejiang Province (Taizhou, China). The SEM analysis of this sample is shown in Figure 1. Figure 1a indicates that the overall particle size distribution of the magnetic seeds is broad, with a minimum particle size of 6.41 μm and a maximum particle size of 244.20 μm, demonstrating a considerable size variation. Figure 1b shows the presence of a few impurities in the magnetic seeds (as indicated by the dark particles at Point 1), which are distinctly different from the magnetite (as indicated by the light particles at Point 2). Figure 1c reveals that the impurities at Point 1 contain other excessive elements, such as Mg, Al, Si, Ca, and Ti, in addition to the major elements Fe and O. The presence of these elements significantly reduces the proportion of Fe, thus weakening the magnetic properties of the particles and increasing the loss rate of the magnetic seeds during subsequent magnetic capture processes. Figure 1d shows that the Fe mass fraction in the magnetite particles at Point 2 is 69.27%; therefore, the theoretical purity of this magnetite is 95.68%, indicating relatively high purity. Figure 2 further provides the energy-dispersive X-ray spectroscopy (EDS) mapping spectra of the raw magnetic seeds, wherein the distributions of Ti and Si are clearly visible. Ti exhibits a distinct granular distribution, while Si presents a scattered point distribution. Combined with the distribution of the major elements Fe and O, the impurities in these magnetic seeds can be presumed to exist in the form of ilmenite monomers and silicate mineral intergrowths. Compared with magnetite, ilmenite and silicate minerals have a relatively lower magnetic susceptibility. The presence of these weakly magnetic impurity mineral phases is the primary reason for the high loss rate of industrial magnetite magnetic seeds during their reuse.

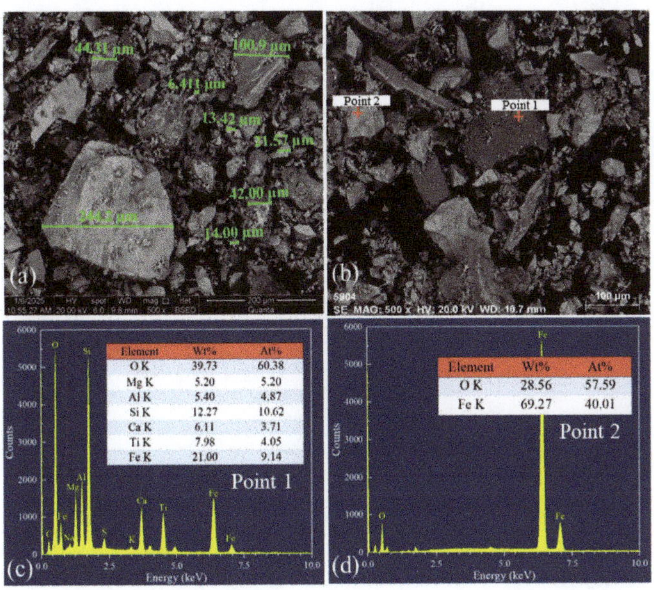

Figure 1. SEM images and EDS spectra of the raw magnetic seeds: (**a**) particle size distribution, (**b**) impurity distribution, (**c**) spectra of Point 1, and (**d**) spectra of Point 2.

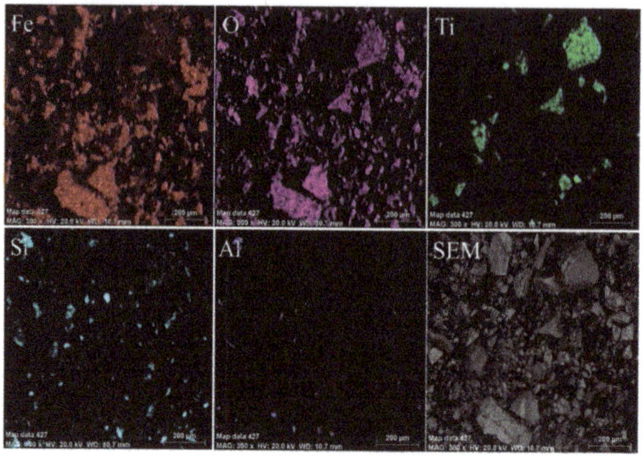

Figure 2. EDS mapping spectra of the raw magnetic seeds.

2.2. Drum Magnetic Separator and Separation Principle

For the magnetic seed recovery equipment, we used a drum-type magnetic separator developed by our team. Its 3D physical drawing, structure diagram, and on-site application are depicted in Figure 3. As seen in Figure 3b, this magnetic separator consists of several major components, including a magnetic system, a rotating roller, a slurry inlet, a sludge outlet, and a magnetic seed discharge area. The magnetic system uses NdFeB permanent magnets (N42), which are arranged in an alternating N–S–S–N magnetic pole structure. The flow channels are designed with both smooth and humped types. The diameter of the roller is 600 mm, and the depth of the flow channel is 30 mm. The humped-type channel features three evenly spaced humps on the bottom wall, as shown in the zoomed view of the flow channel in Figure 3b.

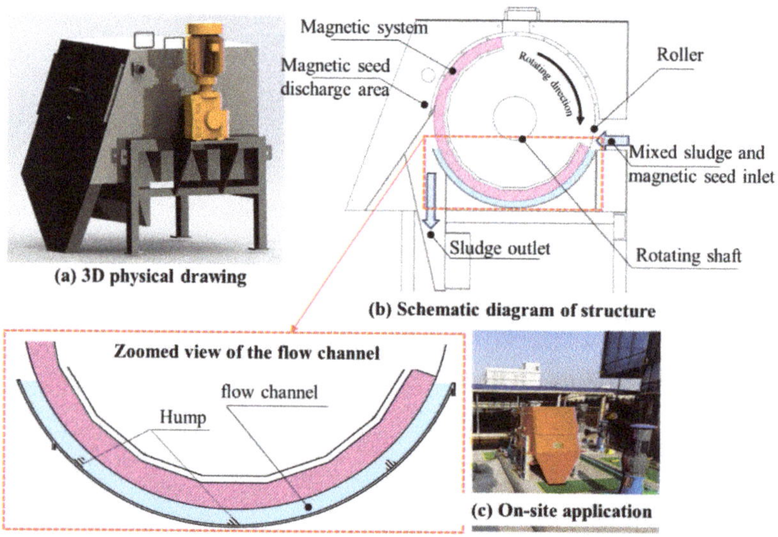

Figure 3. Three-dimensional physical drawing (**a**), schematic of structure (**b**), and on-site application (**c**) of the drum magnetic separator.

The difference in motion trajectories of the magnetic seeds and sludge within the flow channel enables their separation and the reuse of the magnetic seeds. The motion trajectory of particles of different types is primarily determined by the combined body forces acting on particles, such as magnetic, fluid drag, gravitational, centrifugal, and frictional forces, as shown in Figure 4. When the magnetic seeds and sludge flow through the channel of the magnetic separator, the magnetic force acting on the magnetic seed particles must be greater than the combined force of the competing forces (e.g., gravity and fluid drag) [15], causing the particles to be stuck onto the roller surface and then collected in the magnetic seed discharge area for reuse after the drum's rotation. By contrast, the magnetic force acting on the sludge particles is smaller than the combined forces of gravity and fluid drag, causing the sludge particles to flow out through the sludge outlet with the fluid. Particles that have already become stuck onto the roller surface are also subject to centrifugal and frictional forces in addition to magnetic, fluid drag, and gravitational forces due to the drum's rotation. This situation can continue to clean the sludge carried with the magnetic seeds from the roller surface. In practice, researchers frequently assume microfine particles to be spherical to simplify the calculation of body forces [18]. The corresponding magnetic force F_m, fluid drag force F_d, gravitational force F_g, centrifugal force F_c, and frictional force F_f can be calculated using Equations (1)–(5) [19].

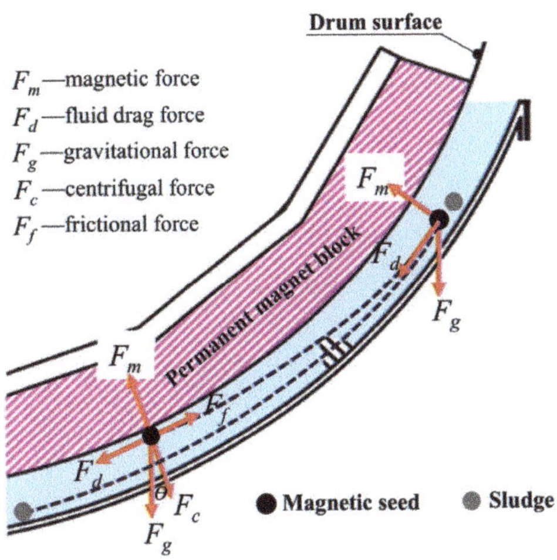

Figure 4. Separation process of the magnetic seeds and sludge in the drum magnetic separator.

The magnetic force acting on particles can be expressed as follows:

$$F_m = \frac{4}{3}\pi R^3 \frac{\kappa B gradB}{\mu_0}, \tag{1}$$

where R is the particle radius, μ_0 is the vacuum permeability, κ is the volume susceptibility, and B and $gradB$ are the induced magnetic field strength and gradient, respectively. The units and specific values of all parameters in this study's equations are provided in Tables 1 and 2.

Table 1. Specific magnitudes of parameters in the body forces.

Parameters	Units	Magnitudes
Particle diameter ($2R$)	m	5–250 $\times 10^{-6}$
Vacuum permeability (μ_0)	N/A^2 or H/m	$4\pi \times 10^{-7}$
Induced magnetic field force ($BgradB$)	$kg^2/(m{\cdot}s^4{\cdot}A^2)$	0–11
Magnetic susceptibility of impurity (κ)	dimensionless	1.0
Viscosity of fluid (η)	Pa·s	1.01×10^{-3}
Flow velocity in the channel (v)	m/s	0–1.31
Density of impurity (ρ_P)	kg/m^3	5.0×10^3
Density of fluid (ρ_f)	kg/m^3	1.0×10^3
Radius of roller (a)	m	0.6
Rotational angular velocity of roller (ω)	rad/s	0.53π
Acceleration due to gravity (g)	m/s^2	9.8

Table 2. Specific magnitudes of parameters in the discrete-element particle trajectory simulation [20].

Parameters	Units	Magnitudes
Particle diameter ($2R$)	m	38×10^{-6}
Density of magnetite/impurity (ρ_P)	kg/m^3	5.0×10^3
Rotational angular velocity of roller (ω)	rad/s	0.53π
Acceleration due to gravity (g)	m/s^2	9.8
Surface energy of magnetite/impurity (γ)	J/m^3	0.05
Shear modulus of magnetite/impurity (G)	Pa	1×10^9
Young's modulus of magnetite/impurity (E)	Pa	2.5×10^9
Magnetic susceptibility of magnetite (κ)	dimensionless	4.0
Magnetic susceptibility of impurity (κ)	dimensionless	1.0
Poisson's ratio of magnetite/impurity (v)	dimensionless	0.25
Restitution coefficient of magnetite/impurity (e)	dimensionless	0.15
Static friction coefficient of magnetite/impurity (f_s)	dimensionless	0.68
Rolling friction coefficient of magnetite/impurity (f_r)	dimensionless	0.05

In addition, the fluid drag, gravitational, and centrifugal forces acting on the particles are, respectively, given as follows:

$$F_d = 6\pi\eta R\Delta v, \tag{2}$$

$$F_g = \frac{4}{3}\pi R^3\left(\rho_P - \rho_f\right)g, \tag{3}$$

$$F_c = \frac{4}{3}\pi R^3\rho_P\omega^2 a, \tag{4}$$

where η is the viscosity of air, Δv is the relative velocity between a particle and airflow, ρ_P is the density of a particle, ρ_f is the density of fluid, ω is the rotational angular velocity of the roller, a is the radius of the roller, and g is the acceleration due to gravity. Friction is the force acting in the reverse direction of particle movement, which is a function of normal force and the coefficient of friction (f). It can be expressed as follows:

$$F_f = f\left(F_m - F_c - F_g \cos\theta\right). \quad (5)$$

2.3. Experimental Methods

This experiment primarily investigated the effect of variations in flow channel shape on the capture of magnetic seeds. Therefore, we disregarded the effect of sludge on magnetic seed capture during the experimental process. In the experiment, 10 L of clean water was first added to a stirring tank, followed by the addition of 1000 g of the raw magnetic seeds. Then, the water and the seeds were thoroughly mixed. The flow channel of the drum magnetic separator was then filled with clean water, and the drum speed was adjusted to set values (12, 16, 24, and 32 r/min). Thereafter, a pump was used to feed the prepared slurry from the stirring tank into the magnetic separator at the desired flow rates (20, 30, 40, and 50 m^3/h) for each smooth and humped channel. Once the slurry had been fully fed, the magnetic seeds that escaped with the clean water were collected. The corresponding mass was recorded after drying and weighing. Subsequently, representative samples of the raw and escaped magnetic seeds were selected for comparative analysis by using SEM and VSM to identify the causes of magnetic seed loss. The method for calculating the recovery rate of the magnetic seeds for different flow channels was as follows:

$$\varepsilon = \left(1 - \frac{M_e}{M_T}\right) \times 100\%, \quad (6)$$

where M_e is the mass of the escaped magnetic seeds, and M_T is the total mass of the feeding magnetic seeds.

3. Experimental Results

3.1. Experimental Conditions

In the comparison experiment of feeding slurry quantity, the rotating speed of the roller was fixed at 16 r/min, and slurry quantity was adjusted separately in the smooth and humped channels. The experimental results are presented in Figure 5. As shown in the figure, the recovery of the magnetic seeds in both channels slowly decreased within the range of 20 m^3/h to 30 m^3/h with the feeding slurry quantity increasing from 20 m^3/h to 50 m^3/h, and then rapidly decreased when the quantity exceeded 30 m^3/h. In particular, recovery dropped from the initial 99.78% to 97.90% in the smooth channel, with a significant decrease. The increase in quantity raised the superficial velocity in the channel, increasing the drag force of the fluid, which affected the capture of the magnetic seeds. Compared with that in the smooth channel, magnetic seed recovery in the humped channel was significantly higher, with the advantage becoming more pronounced as the quantity increased. At quantities of 30 m^3/h and 50 m^3/h, the recovery rates of magnetic seeds in the humped channel were 99.90% and 99.05%, respectively, which were 0.22% and 1.15% higher than those in the smooth channel. The humped channel can effectively mitigate the effect of flow fluctuations on magnetic seed capture during the wastewater treatment process.

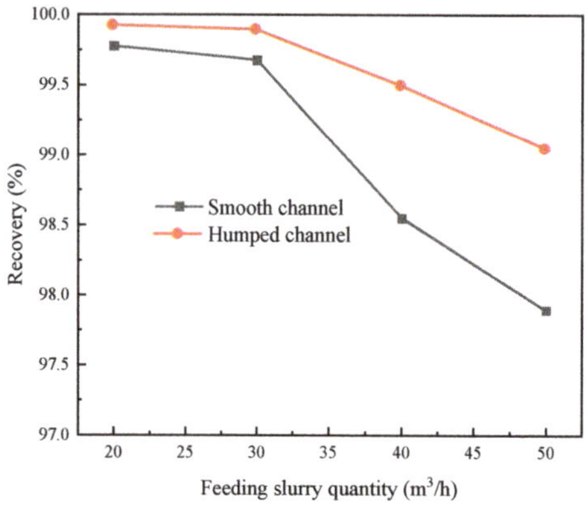

Figure 5. Experimental results of different feeding slurry quantities (with a roller rotating speed of 16 r/min).

In the comparison experiment under varying roller rotating speed conditions, rotating speed was adjusted in the smooth and humped channels with a fixed feeding slurry quantity of 30 m^3/h. The experimental results are presented in Figure 6. As can be observed in the figure, the recovery of the magnetic seeds in both channels consistently decreased as the rotating speed increased from 12 r/min to 32 r/min, with a more pronounced decline at higher speeds. In particular, the recovery rate in the smooth channel decreased from an initial 99.93% to 94.59%, i.e., a reduction of 5.34%. The increase in rotating speed leads to a higher centrifugal force acting on the particles within the channel, thus intensifying the tendency of the particles to move toward the bottom wall of the channel and affecting the capture of the magnetic seeds. Compared with that in the smooth channel, magnetic seed recovery in the humped channel was significantly higher, and the advantage became more evident with higher rotating speeds. At a rotating speed of 16 r/min and 32 r/min, recovery rates in the humped channel were 99.90% and 97.59%, respectively, which were 0.22% and 3.00% higher than those in the smooth channel. Therefore, the humped channel allows for a moderate increase in roller rotating speed without compromising the magnetic seed capture, thus enhancing the processing capacity of the drum magnetic separator.

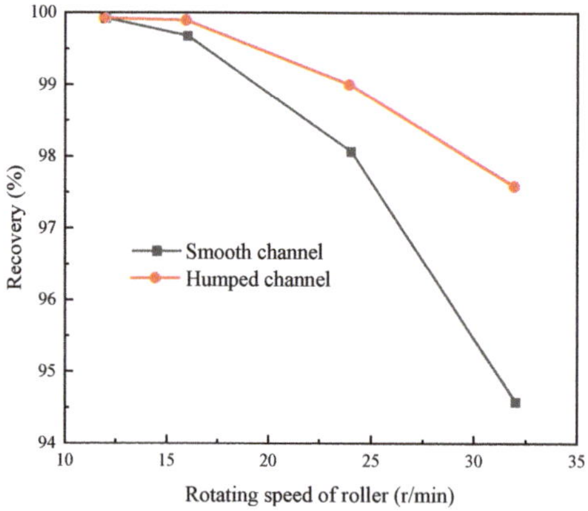

Figure 6. Experimental results of different roller rotating speeds (with a feeding slurry quantity of 30 m^3/h).

3.2. Product Analysis

SEM and VSM analyses of the raw and escaped magnetic seeds were conducted under the following conditions: feeding slurry quantity of 30 m^3/h and roller rotating speed of 16 r/min. The results are presented in Figures 7–9. As depicted in Figure 7, the escaped magnetic seeds are finer in size compared with the feeding raw magnetic seeds. This discrepancy is primarily due to the fact that the magnetic force acting on the particles is proportional to the cube of particle size. That is, smaller particles experience a weaker magnetic force, making them less likely to be stuck onto the roller surface. Further observation of the particle size distribution of the escaped magnetic seeds under the smooth and humped channels reveals that the escaped magnetic seeds exhibit a wide range of sizes in the smooth channel, with some larger particles (>100 μm) still escaping. By contrast, particle size distribution in the humped channel is more uniform, with no significant presence of large particles. This finding suggests that the humped channel increases the capture probability of large magnetic seeds and effectively prevents their loss due to imbalances between magnetic and competing forces (e.g., gravitational and fluid drag forces).

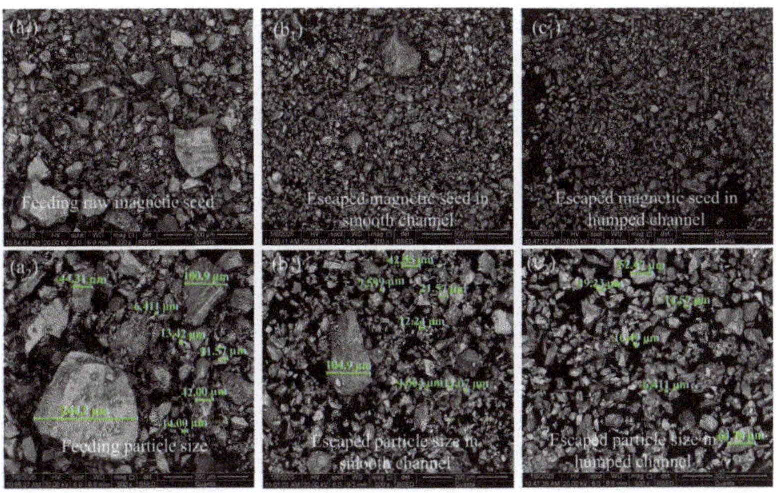

Figure 7. SEM images of the raw magnetic seeds (**a**), escaped magnetic seeds in the smooth channel (**b**), and escaped magnetic seeds in the humped channel (**c**).

Figure 8 further illustrates the EDS mapping spectra of the samples. Based on this figure, the content distribution of Al, Ti, and Si elements in the escaped magnetic seeds is significantly higher than that in the feed magnetic seeds, suggesting that the escaped magnetic seeds are primarily particles with higher impurity content and relatively weaker magnetic properties, making them more difficult to capture. Furthermore, the content of Al, Ti, and Si elements in the escaped magnetic seeds from the humped channel is generally higher than that in escaped magnetic seeds from the smooth channel, indicating that the humped channel enhances the capture of fine, high-purity magnetite particles, thus alleviating the loss of effective magnetic seeds.

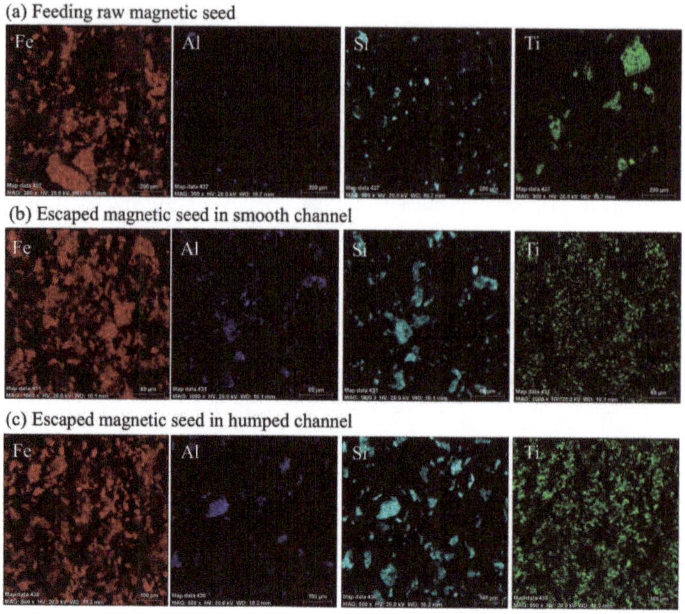

Figure 8. EDS mapping spectra of the raw magnetic seeds (**a**), escaped magnetic seeds in the smooth channel (**b**), and escaped magnetic seeds in the humped channel (**c**).

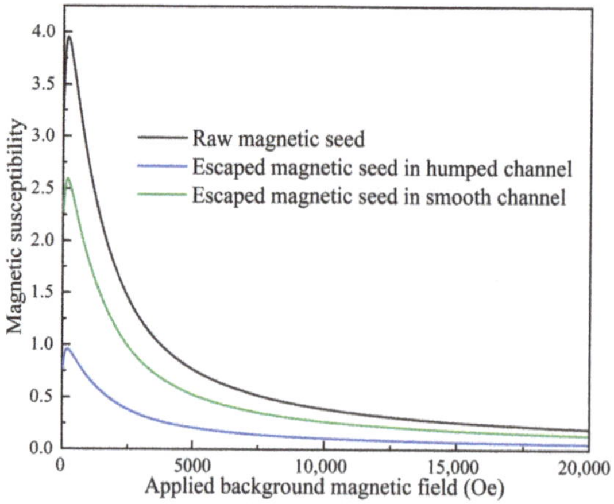

Figure 9. Volume magnetic susceptibility of the feeding and escaped magnetic seeds.

To further investigate the effect of impurity elements, such as Al, Si, and Ti, on the magnetic properties of the magnetic seeds, Figure 9 presents the magnetic susceptibility curves of the feeding and escaped magnetic seeds. Evidently, the susceptibility of the escaped magnetic seeds, which contain higher levels of impurity elements, is significantly lower than that of the feed magnetic seeds. In addition, the susceptibility of the escaped magnetic seeds in the humped channel is lower than that of those in the smooth channel. Therefore, based on the aforementioned results, a conclusion can be drawn that the impurity elements Al, Si, and Ti present in industrial magnetite magnetic seeds affect the purity of certain magnetite particles, reducing their magnetic properties and consequently leading to a higher magnetic seed loss rate. The use of the humped channel in the drum magnetic separator, replacing the traditional smooth channel, helps increase the capture of magnetic seeds and reduce the loss of weakly magnetic impurity seeds.

4. Analysis of Mechanism

For this analysis, we performed a theoretical study on smooth and humped channels by using methods such as finite-element magnetic field simulation [21], computational fluid dynamics simulation [22], numerical estimation of particle body forces, and discrete-element particle trajectory simulation [20], providing a detailed explanation of the enhancement mechanism of magnetic seed capture by the humped channel.

4.1. Simulation of Magnetic and Flow Fields

The magnetic field distribution obtained via finite-element magnetic field simulation is shown in Figures 10 and 11. Figure 10 presents the distribution of the induced magnetic field strength around the magnetic system of the drum magnetic separator, while Figure 11 illustrates the magnitude curves of the induced magnetic field strength and magnetic field force at different distances from the roller surface. The dashed arcs in Figure 10 represent the positions of 1, 14, and 28 mm away from the surface, in which the total depth of the flow channel is 30 mm. From Figure 10, the induced magnetic field strength in the flow channel decreases gradually in the radial direction. In the region within 14 mm of the roller surface, the induced magnetic field strength is consistently greater than 0.15 T, but the induced magnetic field strength is approximately 0.05 T, which is relatively low, near the bottom wall of the channel. The specific magnitude of the induced magnetic field strength at different dashed arcs is shown in Figure 11. Along the circumferential direction of the magnetic system, the induced magnetic field strength and the magnetic field force exhibit a distinct alternating pattern of peaks and valleys, with more pronounced fluctuations closer to the roller surface. At a distance of 1 mm from the roller surface, the maximum induced magnetic field strength and the magnetic field force reach 0.39 T and 11.0 $kg^2/(m{\cdot}s^4{\cdot}A^2)$, respectively. At 14 mm, these values are reduced to 0.21 T and 2.20 $kg^2/(m{\cdot}s^4{\cdot}A^2)$, while at 28 mm, they are further reduced to 0.12 T and 0.60 $kg^2/(m{\cdot}s^4{\cdot}A^2)$, respectively. The reduction in the induced magnetic field strength and the magnetic field force is significant, indicating that magnetic seed particles closer to the roller surface experience stronger magnetic forces.

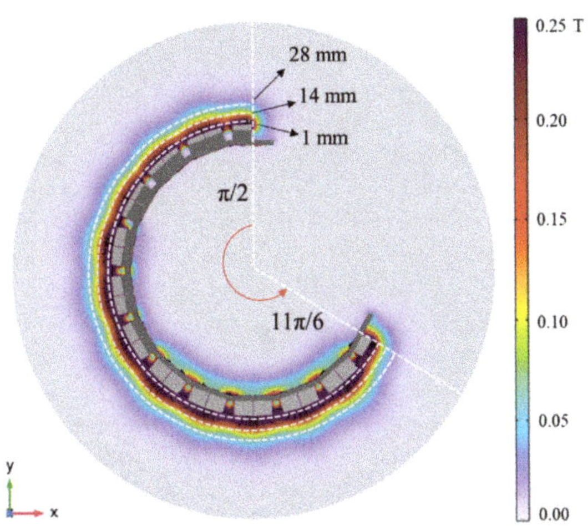

Figure 10. Induced magnetic field distribution of the permanent magnetic system of the drum magnetic separator.

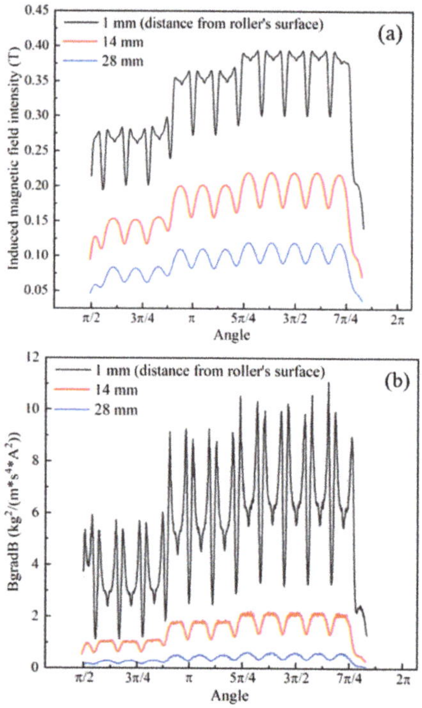

Figure 11. Magnitude curves of the induced magnetic field strength (**a**) and the magnetic field force (**b**) at different distances from the roller surface.

The flow field distributions in the smooth and humped channels, which are obtained through computational fluid dynamics simulation, are shown in Figures 12 and 13, respectively. Figure 12 presents the flow field contour map and vector distribution, while Figure 13 shows the magnitude curves of flow velocity at different heights from the roller surface. As seen in Figure 12a, velocity distribution in the smooth channel is relatively uniform. Overall flow velocity in the channel fluctuates around a superficial velocity of 0.6 m/s, with the deviation being less than 0.1 m/s. In this channel, the magnetic particles move in a relatively regular circumferential motion along the streamlines after entering the flow path. The magnetic particles at the bottom of the channel are unable to transition from regions of low magnetic field force to regions of high magnetic field force, increasing the loss rate. By comparison, the presence of humps disrupts the original regular circumferential flow in the humped channel shown in Figure 12b. At each hump, the channel narrows, and the fluid at the bottom moves upward along the hump, providing kinetic energy for the transition of particles in the magnetic field force, and potentially increasing the probability of magnetic particle capture. The specific flow velocity magnitudes at 1, 14, and 28 mm away from the roller surface in the smooth and humped channels are shown in Figure 13a,b, respectively. As can be observed along the circumferential direction, although the introduction of humps facilitates the transition of magnetic particles toward higher-magnetic-force regions, it also leads to localized velocity surges. Compared to the smooth channel with velocity fluctuations between 0.45 m/s and 0.60 m/s, the maximum velocity in the humped channel can reach 1.14 m/s. Whether it enhances magnetic seed capture still requires further evaluation through the estimation of body forces acting on the particles.

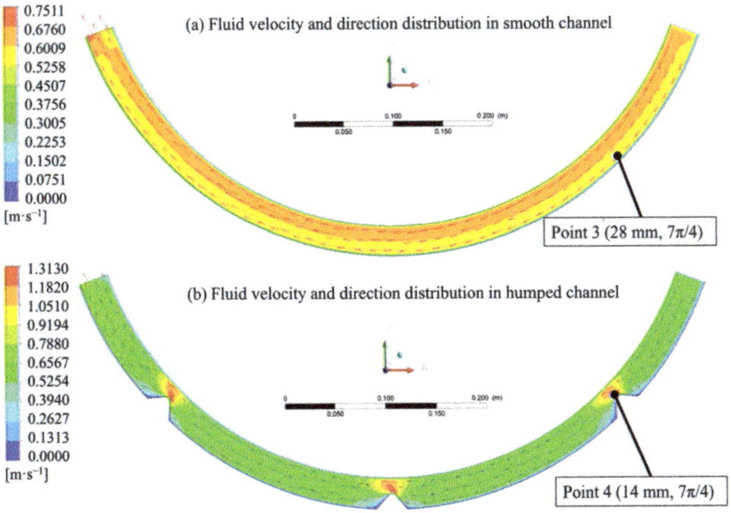

Figure 12. Flow field contour map and vector distribution in the smooth channel (**a**) and the humped channel (**b**).

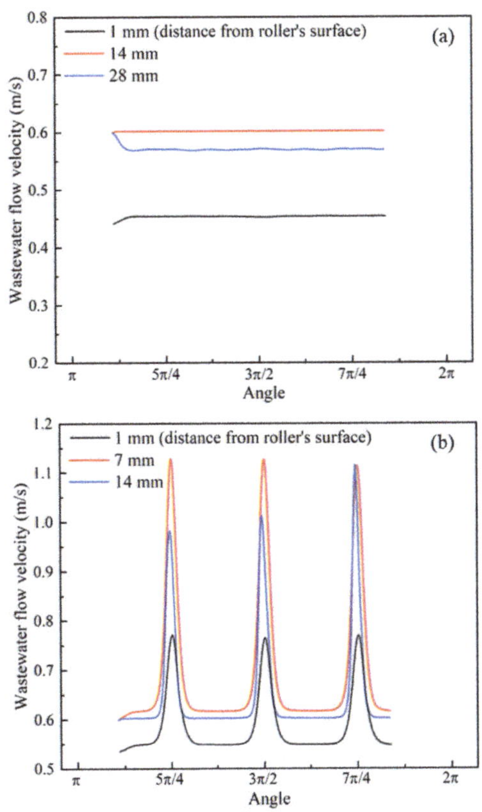

Figure 13. Magnitude curves of flow velocity at different distances from the roller surface in the smooth channel (**a**) and the humped channel (**b**).

4.2. Estimation of Body Forces Acting on Particles

Variation in the shape of the flow channel affects the magnetic force and fluid drag acting on the impure magnetic seeds, while body forces, such as gravitational, centrifugal, and friction forces, remain unchanged. Therefore, only the magnetic and fluid drag forces at Points 3 and 4, as indicated in Figure 12, were calculated using Equations (1) and (2). The specific magnitudes of parameters in these equations are listed in Table 1, and the results are presented in Figure 14. As seen in Figure 14a, the magnetic and fluid drag forces at Point 3 (in the smooth channel) for particles of different sizes are smaller than those at Point 4 (in the humped channel). For particles with a size of 38 μm, the magnetic and fluid drag forces at Point 3 are 1.37×10^{-8} N and 2.06×10^{-8} N, respectively, with the magnetic force being smaller than the drag force, causing the impure magnetic seeds to tend to escape. By contrast, the magnetic and fluid drag forces are 4.57×10^{-8} N and 4.05×10^{-8} N, respectively, at Point 4, with the magnetic force exceeding the drag force, causing the impure magnetic seeds to be captured. Figure 14b further presents the ratio of the magnetic force to the drag force at Points 3 and 4 for particles of different sizes. As shown in the figure, although the humped flow channel leads to a local increase in flow velocity, the ratio of the magnetic force to the drag force in the high-velocity region is significantly increased, and the coarser the particle size, the more pronounced the increase. Therefore, the particle transition phenomenon caused by the humped channel contributes to enhancing the magnetic capture process.

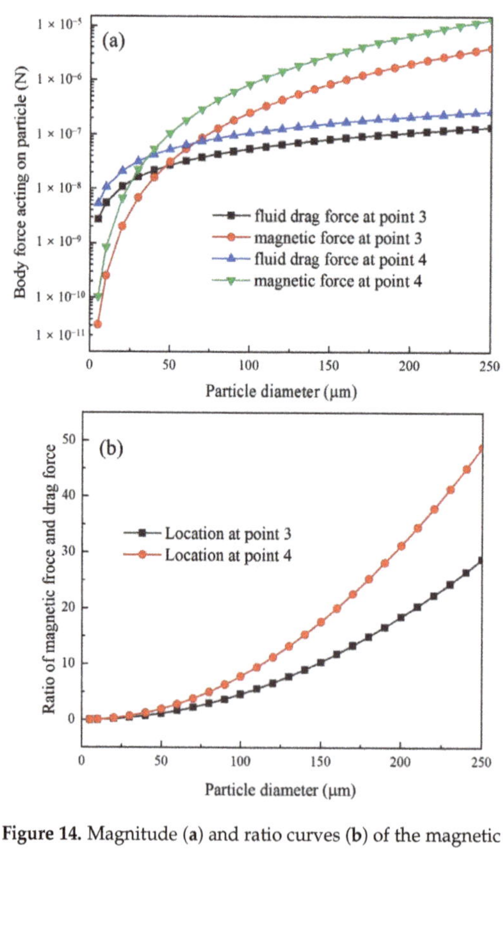

Figure 14. Magnitude (**a**) and ratio curves (**b**) of the magnetic and fluid drag forces at Points 3 and 4.

4.3. Trajectory of Particles

We coupled the simulation data of the magnetic and flow fields with discrete particle motion simulations to obtain the dynamic capture process of magnetite and the impure magnetic seeds in the smooth and humped channels. The specific parameters for particle properties are provided in Table 2, and the computational results are presented in Figure 15 (the dynamic videos are available in the Supplementary Materials). In this figure, black particles represent magnetite, while red particles represent impurities. As seen in Figure 15, the magnetite particles are generally captured and stuck onto the surface of the roller regardless of whether the channel is smooth or humped. However, a significant difference exists in the capture of impurity particles between the two types of channels. In the smooth channel, a portion of the impurity particles is clearly dispersed in the lower region of the channel after 0.8 s and cannot be stuck onto the roller surface. By contrast, the impurity particles noticeably shrink toward the roller surface after passing through the first hump in the humped channel. After 0.8 s, they do not disperse to the bottom of the channel but remain on the roller surface, significantly improving the capture rate. This result further theoretically confirms that the humped channel effectively alleviates the issue of excessive magnetic seed loss due to the presence of impurities during the industrial recycling process of magnetic seeds.

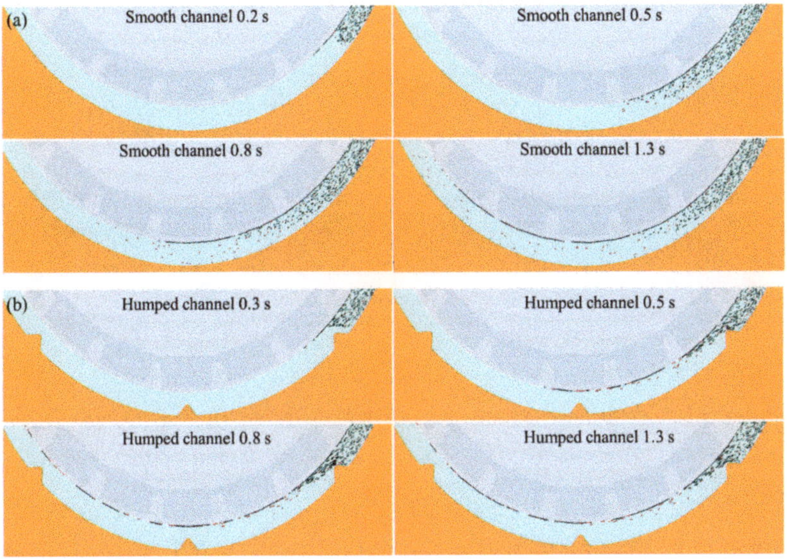

Figure 15. Dynamic capture process of magnetite (black) and impurity (red) magnetic seeds in the smooth (**a**) and humped (**b**) channels.

5. Conclusions

This study aimed to address the issue of high loss rates of industrial magnetic seeds in the magnetic flocculation process. The effect of smooth and humped channels on magnetic seed recovery was studied in a drum magnetic separator. SEM and VSM were utilized to analyze separated samples, identify the primary causes of high magnetic seed loss, and confirm the enhanced recovery mechanism of the humped channel through magnetic field simulation, flow field simulation, particle force calculation, and discrete-element coupling particle trajectory simulation. The major conclusions drawn are as follows:

(1) The SEM and VSM analyses of the samples indicate that a small amount of impurities, which are rich in elements such as Al, Si, and Ti, exist in the industrial magnetite magnetic particles. These impurities have lower magnetization, resulting in lower capture rates in the conventional smooth-channel drum magnetic separator and contributing to the high rate of magnetic seed loss.

(2) The experimental results of magnetic seed capture in channels of different shapes indicate that the recovery of magnetic seeds in the humped channel consistently outperforms that in the smooth channel under various feeding flow rates and drum rotational speeds, with a recovery rate of up to 3%. The humped channel alleviates the high magnetic seed loss problem.

(3) The magnetic field, flow field, and particle trajectory analysis results show that the magnetic field force at the bottom of the smooth channel is only 0.6 $kg^2/(m{\cdot}s^4{\cdot}A^2)$, which is around 18 times lower than that at the roller surface. The application of the humped channel pushes the impure magnetic seeds to transition from the region with a low magnetic field force to a region with high magnetic field force, significantly improving the rate of impure magnetic seed capture.

Supplementary Materials: The following supporting information can be downloaded at: https://www.mdpi.com/article/10.3390/min15070732/s1, Video S1: Dynamic capture process of magnetite (black) and impurity (red) magnetic seeds in the smooth channel; Video S2: Dynamic capture process of magnetite (black) and impurity (red) magnetic seeds in the humped channel.

Author Contributions: Investigation, data curation, formal analysis, writing—original draft preparation, S.X.; conceptualization, supervision, methodology, writing—review and editing, H.H.; analysis, methodology, J.L.; supervision, writing—review and editing, W.S.; methodology, software, J.Q. All authors have read and agreed to the published version of the manuscript.

Funding: This research was financially supported by the Natural Science Foundation of Guangdong Province (Grant No. 2022A1515010754) and the Science and Technology Project of Meizhou (No. 2020B0204002).

Data Availability Statement: Data are available upon request from the authors.

Conflicts of Interest: The authors declare no conflicts of interest.

References

1. Sun, Y. Research progress of magnetic flocculation in water treatment. *Magnetochemistry* **2024**, *10*, 56. [CrossRef]
2. Ellis, E.P.; Cathcart, A.H. Selection, installation, startup and testing of the world's first full-scale comag phosphorus reductio tertiary treatment system. *Proc. Water Environ. Fed.* **2008**, *13*, 3602–3621. [CrossRef]
3. Mohamed Noor, M.H.; Wong, S.; Ngadi, N.; Mohammed Inuwa, I.; Opotu, L.A. Assessing the effectiveness of magneti nanoparticles coagulation/flocculation in water treatment: A systematic literature review. *Int. J. Environ. Sci. Technol.* **2022**, *1* 6935–6956. [CrossRef]
4. Hu, J.W.; Li, J.; Chen, Y.; Chen, X.L. Application of magnetic flocculation technology in enhanced treatment of polluted rive water. *China Water Wastewater* **2011**, *27*, 1000–4602. (In Chinese)
5. Xie, C.; Li, X.; Tang, Z.; Chu, W.; Li, H. Enhanced combined sewer overflow treatment by rapid magnetic flocculation-magneti sedimentation: Efficiency and mechanism. *Environ. Sci. Water Res. Technol.* **2024**, *10*, 1586–1594. [CrossRef]
6. Zhou, Q.; Yu, Z.; Ma, Y. Review on the application of magnetic flocculation technology in water treatment. *IOP Conf. Ser. Eart Environ. Sci.* **2019**, *295*, 042107. [CrossRef]
7. Li, Y.; Wang, J.; Zhao, Y.; Luan, Z. Research on magnetic seeding flocculation for arsenic removal by superconducting magneti separation. *Sep. Purif. Technol.* **2010**, *73*, 264–270. [CrossRef]
8. Du, C.; Hu, Y.; Han, H.; Sun, W.; Hou, P.; Liu, R.; Wang, L.; Yang, Y.; Liu, R.; Sun, L.; et al. Magnetic separation of phosphat contaminants from starch wastewater using magnetic seeding. *Sci. Total Environ.* **2019**, *695*, 133723. [CrossRef]
9. Zhang, L.; Wang, B.; Zhang, Q. Enhanced treatment of oily wastewater through modified magnetic seeds for magnetic flocculatior *Int. J. Environ. Sci. Technol.* **2024**, *21*, 7717–7732. [CrossRef]

Wang, X.X.; Wu, L.; Zhou, P.; Li, C.; Zhao, L.B.; An, W.; Chen, Y. Harvesting of polycyclic aromatic hydrocarbons (pahs)-degrading bacteria by magnetic flocculation. *Appl. Mech. Mater.* **2014**, *665*, 563–566. [CrossRef]

Yiacoumi, S.; Rountree, D.A.; Tsouris, C. Mechanism of particle flocculation by magnetic seeding. *J. Colloid Interface Sci.* **1996**, *184*, 477. [CrossRef]

Kurinobu, S.; Uesugi, J. Performance of hgms filter and recycling of magnetic seeding material on magnetic seeding method. *IEEE Trans. Magn.* **1999**, *35*, 4067–4069. [CrossRef]

Du, H.; Zhang, F.; Lv, F.M. Experimental study on magnetic flocculation-magnetic disc separation technique for chaohu eutrophication of water treatment. *Technol. Water Treat.* **2009**, *35*, 86–90.

Arol, A.I.; Aydogan, A. Recovery enhancement of magnetite fines in magnetic separation. *Colloids Surf. A Physicochem. Eng. Asp.* **2004**, *232*, 151–154. [CrossRef]

Tang, D.; Wang, F.; Dai, H.; Lu, M.; Gong, Z. Influence of separation chamber shape in dry magnetic separator on the dispersion and separation of multiple magnetites. *Miner. Eng.* **2021**, *171*, 107130. [CrossRef]

Wang, F.; Tang, D.; Dai, H.; Zhang, S.; Gao, L.; Gong, M. Design optimization and manufacturing of an innovative precise low-intensity magnetic separator based on a multiphysics model. *Miner. Eng.* **2021**, *171*, 107118. [CrossRef]

Guo, Y.F.; Dong, H.G.; Jiang, T. Recovering magnetite from nickel metallurgical residue by elective flocculation-magnetic separation. *J. Cent. South Univ. Sci. Technol.* **2006**, *37*, 680–684.

Hu, Z.; Lu, D.; Zheng, X.; Wang, Y.; Xue, Z.; Xu, S. Development of a high-gradient magnetic separator for enhancing selective separation: A review. *Powder Technol.* **2023**, *421*, 118435. [CrossRef]

Hu, Z.; Lu, D.; Wang, Y.; Zheng, X.; Zhang, Y. A novel pneumatic dry high-intensity magnetic separator for the beneficiation of fine-grained hematite. *Powder Technol.* **2024**, *433*, 119216. [CrossRef]

Li, X. Equipment and Fundamental Research of Dry Magnetic Separation for Micro Fine Particles Enhanced by Airflow. Ph.D. Thesis, Central South University, Changsha, China, 2023. (In Chinese).

Hu, Z.; Liu, J.; Gan, T.; Lu, D.; Wang, Y.; Zheng, X. High-intensity magnetic separation for recovery of LiFePO4 and graphite from spent lithium-ion batteries. *Sep. Purif. Technol.* **2022**, *297*, 121486. [CrossRef]

Yi, F.; Chen, L.; Zeng, J. Combinatorial optimization of rotating matrix in centrifugal high gradient magnetic separation. *Miner. Eng.* **2023**, *202*, 108309. [CrossRef]

minerals

MDPI

Article

The Optimized Synthesis of Barium Sulfate: A Scalable and Sustainable Laboratory Approach Using D-Optimal Design

Abdulkarim Shafiee [1], Mohammad Sadatipour [1], Fatemeh Sadat Hoseinian [1,*], Bahram Rezai [1] and Mehdi Safari [2,3,*]

1 Department of Mining Engineering, Amirkabir University of Technology, Tehran 158754413, Iran
2 Minerals Processing Division, Mintek, Randburg 2125, South Africa
3 Faculty of Engineering and the Built Environment, University of the Witwatersrand, 1 Jan Smuts Ave., Johannesburg 2000, South Africa
* Correspondence: f_hoseinian@aut.ac.ir (F.S.H.); mehdis@mintek.co.za (M.S.)

Abstract: This study introduces a controlled laboratory-scale method for synthesizing barium sulfate, with promising potential for industrial scalability to meet increasing demand for quality and sustainable production. Statistical tools—specifically, design of experiments (DOE) and the D-optimal methodology via DX12.0.3.0 software—were employed to determine optimal reaction parameters. The synthesis process involved the precise control of barium chloride and sulfate concentrations, feed rate, and temperature to ensure the production of barium sulfate. Under optimized conditions ($BaCl_2$ = 1.5 M, SO_4^{2-} = 0.41 M, flow rate = 2.53 mL/min, temperature = 20.03 °C), a yield of 9.20 g of barium sulfate with a density of 4.25 g/cm^3 was achieved. The analysis confirmed the purity of the product, validating its suitability for demanding industrial applications. This approach not only improves product quality but also minimizes waste and reduces operational costs, offering a sustainable and scalable solution for barium sulfate production. These findings mark a meaningful step forward in mineral processing and purification technologies.

Keywords: barium sulfate; optimization; purity; density; precipitation

1. Introduction

Barite ($BaSO_4$) has transitioned from being a low-value mineral to a resource of strategic significance across various industries, especially in the energy sector. Its unique attributes, including high density and chemical inertness, render it essential for oil exploration and drilling. Notably, nearly 90% of global barite consumption is attributed to its function as a weighting agent in drilling fluids, which are critical for stabilizing wellbores and managing formation pressures to ensure operational safety and efficiency [1–6]. The exceptional density and minimal solubility of barite make it ideally suited for drilling applications, as it effectively counters formation fluid pressures, thereby reducing blowout risks and enhancing stability. Beyond oil and gas, barite's versatility has facilitated its adoption across diverse industries. Its physical and chemical properties are leveraged in pigment production, where its high refractive index and chemical inertness contribute to superior optical and functional traits [7,8]. As a critical component in drilling mud, barite's characteristics—being clean, soft, inert, and economically viable—are further amplified by its specific gravity exceeding 4 g/cm^3, making it an effective weighting agent [9,10]. Its functions extend to cooling and lubricating drill bits, lining well casings, and facilitating drilling fluid transport, concurrently controlling abnormal formation pressures [11]. Natural barium sulfate with a purity of 90%–95% and minimal iron oxide contamination (<1%)

is extensively utilized in manufacturing high-quality white pigments and as a viscosity modifier in paints. In contrast, synthetic barium sulfate, commonly known as blanc fixe, is produced through the precipitation of pure barium sulfate from a chemical reaction between sodium sulfate (Glauber's salt) and barium sulfide [12–15]. In pharmaceuticals, although overall industrial mineral consumption is lower, barite's economic value is notable when processed to meet rigorous standards, requiring purity levels exceeding 97.5% [16–18]. Barium sulfate serves a crucial role as a contrast agent in X-ray and CT imaging, enhancing diagnostic image quality by improving the visibility of internal structures [19–21]. However, care is necessary for patients with certain conditions due to the potential adverse effects if used improperly [22,23]. In cases where barite contains witherite, the resulting solution can become toxic, potentially causing kidney dysfunction or paralysis. Therefore, the use of barium sulfate in medical procedures should always adhere to the guidance of a qualified healthcare professional [24–26].

At the turn of the 19th and early 20th centuries, several materials were tested as contrast agents for X-ray imaging. Initially, heavy metal salts and oxides were suggested for this purpose. The first material employed was bismuth nitrogen compounds, which, however, caused significant toxicity. Subsequently, bismuth carbonate and high-atomic-number metal oxides, such as thorium oxides, were utilized. Due to the hazardous nature of these substances, barite (barium sulfate) emerged as a safer alternative. Nevertheless, the use of barite was not devoid of complications, as impurities often resulted in toxic effects [27–29]. In 1911, the Mallinckrodt Chemical Company introduced a method called the "black ash process" to purify barite and convert it into soluble barium salts [30]. By 1913, this method was industrially implemented, marking a significant advancement in barite refinement. The black ash process facilitated the removal of impurities, rendering barite a safer and more efficient material for diverse applications [31–33]. In 1930, Henry Farr proposed an alternative process using calcium chloride instead of oil and gas to convert barium sulfate. This method involved purifying barite and transforming it into a complex salt, followed by hydrolysis of the complex salt, which led to the precipitation of barium sulfate. Other salts, such as potassium chloride, potassium nitrate, and chromium or iron sulfates, could also form complex salts with barite. This method produced fine precipitates, effectively separating impurities like iron oxides and siliceous materials from barite [34–36]. Later, in 1936, Norris and colleagues expanded on Farr's invention and succeeded in extracting soluble salts from both barium and strontium sulfates. The reaction demonstrated optimal efficiency when equimolar ratios of barium sulfate and calcium chloride (containing 3–5 parts water) were subjected to intense heat for a brief period, resulting in their fusion. The resulting product was cooled using a methanol–ethylene glycol mixture, yielding barium chloride with a purity of 95% or higher [37–41]. Tasbolatovna reviewed various methods for producing high-purity barium sulfate. The study highlighted common impurities in the synthesis process, including sodium and ammonium chlorides, as well as sulfur-based compounds such as sulfur chloride or SO_2. The use of barium hydroxide was found to effectively reduce sulfide impurities. In one method, sulfuric acid and hydrochloric acid were employed for washing barite, with high-purity barite being essential for optimal results. This process upgraded barite with an initial purity of 95%–98% to a final grade of 99% purity. Additionally, barite purification was achieved by dissolving it in molten salt mixtures of sodium chloride and calcium chloride. After separating the solid and molten phases and washing with water, pure barium sulfate was obtained. This high-purity barium sulfate has proven applications in industries such as battery production and papermaking [42].

Recent studies emphasize the growing need for sustainable mineral processing routes to produce high-purity compounds with minimal environmental impact [43]. These applications require not only high chemical purity but also narrow particle size distribution and high density, which are difficult to achieve using conventional techniques. Our approach aligns with these modern standards by applying statistically optimized synthesis methods to produce $BaSO_4$ with over 96% purity, reducing reagent waste. In addition to achieving high purity and yield, the proposed synthesis method demonstrates improved environmental sustainability. The use of a statistically optimized D-optimal design allowed for minimal reagent consumption—specifically reducing excess use of barium chloride and sulfuric acid [44–46]. As the reaction proceeds via a simple precipitation mechanism under ambient temperature and pressure, it eliminates the need for high-energy input or hazardous catalysts. Furthermore, no secondary or toxic by-products are formed during the process, and the resulting supernatant is non-corrosive and can be treated using standard neutralization protocols, making the approach compliant with green chemistry principles. Compared to conventional high-temperature or multi-step methods, our process significantly lowers the environmental footprint and aligns with recent efforts to develop cleaner mineral synthesis techniques [47,48].

In this study, we introduce a streamlined and environmentally conscious synthesis method for producing barium sulfate. Traditional synthesis routes often require extensive material input and generate unnecessary waste. In contrast, our methodology leverages the D-optimal design approach within a Design of Experiments (DOE) framework to systematically optimize key variables—including reactant concentrations, flow rate, and reaction temperature. Implemented through DX12 software, this statistical strategy allowed for efficient identification of ideal process conditions using fewer experimental runs [49–51]. By refining these parameters, we achieved consistent product quality with improved mass yield and density, while minimizing chemical usage and operational overhead. This integrated approach not only enhances scalability and reproducibility but also aligns with sustainable processing practices by reducing environmental impact—addressing both performance and ecological considerations in industrial mineral synthesis.

2. Materials and Methods

2.1. Production of Barium Sulfate

Barium carbonate ($BaCO_3$), hydrochloric acid (HCl) and sulfuric acid (H_2SO_4) were used. All reagents used in the experiments are industrial grade purity reagents. Barium chloride ($BaCl_2$) was synthesized by reacting barium carbonate ($BaCO_3$) with hydrochloric acid (HCl) according to the following reaction (Equation (1)):

$$2HCl(aq) + BaCO_3(s) \rightarrow BaCl_2(aq) + H_2O(aq) + CO_2(g) \tag{1}$$

Hydrochloric acid was added to an acid-resistant vessel, followed by the gradual introduction of barium carbonate powder while continuously stirring with a mechanical stirrer. This ensured an enhanced reaction rate and prevented premature precipitation of barium carbonate. The reaction proceeded until the pH was adjusted neutrality, at which point a portion of barium carbonate remained as a precipitate. Once complete sedimentation was achieved, the remaining solution was carefully separated for further processing.

To prepare 1 L of 1 M H_2SO_4, 55 mL of concentrated sulfuric acid was carefully added to approximately 800 mL of distilled water in a fume hood with continuous stirring, and then the solution was diluted to a final volume of 1 L using a volumetric flask. During the experimental procedure, varying concentrations of barium chloride solution were reacted with the prepared H_2SO_4 under magnetic stirring. This reaction resulted in the formation of barium sulfate ($BaSO_4$) as a precipitate and hydrochloric acid (HCl) as a byproduct, as represented by the following chemical equation (Equation (2)):

$$BaCl_2(aq) + H_2SO_4(aq) \rightarrow BaSO_4(s) + 2HCl(aq) \quad (2)$$

The density of the barium sulfate samples was determined using a standard pycnometer method in accordance with ASTM D2330 [52]. A pycnometer of known volume (25 mL) was used for the measurement. Approximately 5–15 g of oven-dried sample was accurately weighed and placed into the pycnometer. Deionized water was then added to fill the pycnometer, ensuring that no air bubbles remained trapped. Then, the mass of the filled pycnometer was recorded. The density of the dry sample (ρ_s) was calculated using the following equation (Equation (3)):

$$\rho_s = \frac{m_s \times \rho_w}{(m_{pw} - m_p) - (m_{psw} - m_p - m_s)} \quad (3)$$

where

m_s is the mass of the dry sample,

m_p is the mass of the empty pycnometer,

m_{pw} is the mass of the pycnometer filled with water,

m_{psw} is the mass of the pycnometer filled with sample and water,

ρ_p is the density of the water at the measurement temperature.

Each measurement was performed in triplicate, and the average value was reported to ensure accuracy and repeatability.

2.2. Experimental Design

The following sections describe the experimental design and synthesis methodology used to produce barium sulfate via a synthetic approach. Initial experiments revealed a direct correlation between the barium concentration and the density of the resulting barium sulfate. Consequently, the density of the synthesized barium sulfate was selected as a key response variable for the optimization study. To efficiently determine the optimal synthesis conditions while minimizing the number of experimental trials, the design of experiments (DOE) methodology was applied. This statistical approach allows for the evaluation of both individual parameters and their interactions on the synthesis process [53–55]. The experimental design was developed using DX12 software, employing the D-optimal design strategy to guide the optimization.

The input parameters were defined as the barium chloride concentrations, sulfate concentrations, feed rate, and temperature. The output parameters were the mass of the produced barium sulfate (in grams) and its density. The minimum and maximum concentrations of barium chloride and sulfuric acid were set between 0.35 M and 1.5 M, based on the solubility product constants (K_{SP}) of the chemical compounds, the relevant literature, and experimental constraints. The concentration range was considered to be from 0.35 to 1.5 M. Increasing the concentration beyond this value would cause high acidity of the medium and failure of the reaction to complete. The feed rate was determined to range from 2.5 to 20 mL/min, considering the limitations of the laboratory burette. The solution temperature range was specified as 20–80 °C. The input parameter ranges used in the software, based on the results of preliminary experiments, are provided in Table 1. Additionally, the experimental design by DX12 software is detailed in Table 2.

Table 1. The range of the input variables used in the experimental design.

Parameter	Type	Minimum Value	Maximum Value	Mean
Barium chloride concentration (mol/L)	A	0.35	1.5	0.919
Sulfate concentration (mol/L)	B	0.35	1.5	0.904
Feed rate (mL/min)	C	2.5	20	10.617
Temperature (°C)	D	20	80	53.776

Table 2. Experimental design for process optimization.

Experiment No.	Barium Chloride Concentration (M)	Sodium Sulfate Concentration (M)	Flow Rate (mL/min)	Temperature (°C)
1	1.5000	0.4398	20.00	72.05
2	0.3500	1.5000	2.50	20.00
3	1.3742	0.7143	11.51	46.57
4	0.3500	0.3500	2.50	80.00
5	0.3500	0.9625	20.00	42.49
6	1.5000	1.5000	20.00	80.00
7	1.5000	1.5000	11.24	39.45
8	1.5000	0.8547	2.50	20.00
9	1.5000	0.8547	2.50	20.00
10	0.9634	0.3500	2.50	42.86
11	0.3500	1.4820	2.50	71.46
12	0.8551	1.5000	20.00	20.00
13	1.5000	0.3500	5.58	80.00
14	1.5000	1.5000	2.50	80.00
15	0.3500	0.3500	20.00	80.00
16	0.3500	1.5000	17.32	80.00
17	0.9191	0.9310	11.21	80.00
18	1.5000	1.5000	2.50	80.00
19	0.9250	0.9250	11.25	20.00
20	0.3500	0.3500	2.50	80.00
21	1.5000	0.3500	20.00	20.00
22	0.3500	0.3500	11.23	20.00
23	0.9483	1.1748	3.59	47.04
24	0.3500	0.9625	20.00	42.49
25	0.3500	0.3500	20.00	80.00

3. Results and Discussion

The product process optimization of barium sulfate was evaluated using DX12 software. In this regard, the key parameters, including barium chloride concentration, sulfate concentration, feed rate, and temperature, were studied.

3.1. Numerical Experimental Results

The input parameters considered were barium chloride concentration, sulfate concentration, feed rate, and temperature, while the output parameters were the mass of the produced barium sulfate and its density. The experimental results were analyzed using DX12 software. The regression equations for the mass of the produced barium sulfate and density are given as Equations (4) and (5), respectively (the models are expressed in terms of coded variables).

$$
\begin{aligned}
\text{Mass} = \; & 15.48 + (4.91 \times \text{A}) + (7.36 \times \text{B}) + (-0.34 \times \text{C} + (-0.10 \times \text{D}) + (4.37 \times \text{AB}) \\
& + (0.19 \times \text{AD}) + (-0.16 \times \text{BC}) + (0.23 \times \text{CD}) + \left(-2.95 \times \text{A}^2\right) + \left(-1.87 \times \text{B}^2\right) \\
& + \left(-0.60 \times \text{C}^2\right) + \left(1.58 \times \text{D}^2\right) + \left(-1.06 \times \text{A}^2\text{B}\right) + \left(-1.60 \times \text{A}^2\text{D}\right) + \left(0.74 \times \text{AD}^2\right) \\
& + \left(-1.79 \times \text{BC}^2\right) + \left(2.45 \times \text{C}^2\text{D}\right)
\end{aligned} \tag{4}
$$

$$
\begin{aligned}
\text{Density} = \; & 3.90 + (0.06 \times \text{A}) + (-0.07 \times \text{B}) + (-0.03 \times \text{C}) + (-0.04 \times \text{D}) + (-0.04 \times \text{AB}) \\
& + (-0.004 \times \text{BC}) + (0.06 \times \text{BD}) + (0.02 \times \text{CD}) + \left(0.03 \times \text{A}^2\right) + \left(-0.05 \times \text{B}^2\right) \\
& + (-0.05 \times \text{BCD}) + \left(0.04 \times \text{B}^2\text{C}\right)
\end{aligned} \tag{5}
$$

The statistical results of the developed models are presented in Table 3. The value of the adequate precision component exceeds four, indicating a satisfactory level of accuracy [56,57]. Other parameters, such as the coefficient of determination (R^2), also demonstrate the efficacy of these models in predicting the mass of the produced barium sulfate and density [58–60].

Table 3. Statistical evaluation of model accuracy in predicting mass and density of synthesized barium sulfate.

Model	R^2	Adjusted R^2	Standard Deviation	Mean	Adequate Precision	Coefficient of Variation (%)
Mass (g)	1.00	0.99	12.86	12.86	247.33	0.78
Density (g/cm^3)	0.82	0.64	0.083	3.88	7.80	2.16

The data presented in Table 4 show the results of the model for the mass of produced barium sulfate. The F-value of the model is 8458.53, which indicates that the model is statistically significant. The probability of obtaining an F-value this large due to random variation or noise is only 0.01%, affirming the robustness of the model. Furthermore, *p*-values less than 0.05 suggest that the corresponding model terms are statistically significant. In this context, the following terms are significant: A, B, C, AB, AD, BC, CD, A^2, B^2, C^2, D^2, A^2B, A^2D, AD^2, BC^2, and C^2D. On the other hand, terms with *p*-values greater than 0.1 are considered not significant, implying that their contribution to the model is negligible. As presented in Table 5, the data correspond to the model of barium sulfate density. The F-value of 4.59 indicates that the model is statistically significant. This value suggests that there is only a 0.66% probability that such an F-value could arise due to random variation or noise, reinforcing the robustness of the model. Further analysis reveals that *p*-values less than 0.0500 denote statistically significant model terms. In this case, the model terms A, B, and BD are significant, confirming their relevance in predicting the barium sulfate density. Conversely, any model term with a *p*-value greater than 0.1000 is considered non-significant, and such terms do not contribute meaningfully to the model's predictive capability. This statistical significance of the model terms and the low probability of noise-driven results underline the model's strong predictive accuracy and reliability [61].

Table 4. Results of the analysis of variance (ANOVA) for the developed model on the mass of barium sulfate produced.

Source	Sum of Squares	df	Mean Square	F-Value	*p*-Value	
Model	1449.19	17	85.25	8458.53	<0.0001	significant
A	22.61	1	22.61	2243.47	<0.0001	
B	22.61	1	22.61	2243.12	<0.0001	
C	1.03	1	1.03	102.42	<0.0001	
D	0.0200	1	0.0200	1.99	0.2013	
AB	94.39	1	94.39	9365.88	<0.0001	
AD	0.1983	1	0.1983	19.68	0.0030	
BC	0.2080	1	0.2080	20.64	0.0027	
CD	0.3143	1	0.3143	31.18	0.0008	
A^2	14.87	1	14.87	1475.61	<0.0001	
B^2	4.10	1	4.10	406.45	<0.0001	
C^2	0.3894	1	0.3894	38.64	0.0004	
D^2	4.84	1	4.84	480.01	<0.0001	
A^2B	0.8433	1	0.8433	83.68	<0.0001	
A^2D	1.36	1	1.36	135.19	<0.0001	
AD^2	0.3161	1	0.3161	31.37	0.0008	
BC^2	2.37	1	2.37	235.36	<0.0001	
C^2D	2.63	1	2.63	261.00	<0.0001	
Residual	0.0705	7	0.0101			
Lack of Fit	0.0055	2	0.0027	0.2113	0.8164	not significant
Pure Error	0.0650	5	0.0130			
Cor Total	1449.26	24				

Table 5. Results of the analysis of variance (ANOVA) for the developed model on the density of barium sulfate produced.

Source	Sum of Squares	df	Mean Square	F-Value	*p*-Value	
Model	0.3864	12	0.0322	4.59	0.0066	significant
A	0.0803	1	0.0803	11.43	0.0055	
B	0.0633	1	0.0633	9.02	0.0110	
C	0.0020	1	0.0020	0.2837	0.6040	
D	0.0199	1	0.0199	2.83	0.1182	
AB	0.0172	1	0.0172	2.45	0.1431	
BC	0.0002	1	0.0002	0.0233	0.8812	
BD	0.0425	1	0.0425	6.06	0.0300	
CD	0.0026	1	0.0026	0.3659	0.5565	
A^2	0.0021	1	0.0021	0.2941	0.5975	
B^2	0.0074	1	0.0074	1.06	0.3243	
BCD	0.0143	1	0.0143	2.03	0.1793	
B^2C	0.0032	1	0.0032	0.4532	0.5136	
Residual	0.0842	12	0.0070			
Lack of Fit	0.0510	7	0.0073	1.09	0.4772	not significant
Pure Error	0.0333	5	0.0067			
Cor Total	0.4707	24				

3.1.1. Mass of Barium Sulfate

Figure 1 displays the normal probability plot for the barium sulfate mass model, illustrating the distribution of residuals and the likelihood of each experimental outcome based on residual interpolation. The clustering of data points along the reference line indicates that the model accurately captures the relationship between sulfuric acid and barium chloride in barium sulfate production. This alignment suggests strong predictive capability and minimal deviation in error distribution. Figure 2 further supports this by comparing the actual mass of barium sulfate produced with the model's predicted values. The data points lie closely along the line of identity, confirming the model's robustness and reliability in predicting barium sulfate mass. Together, these plots validate the model's effectiveness in representing the key dynamics of the synthesis process.

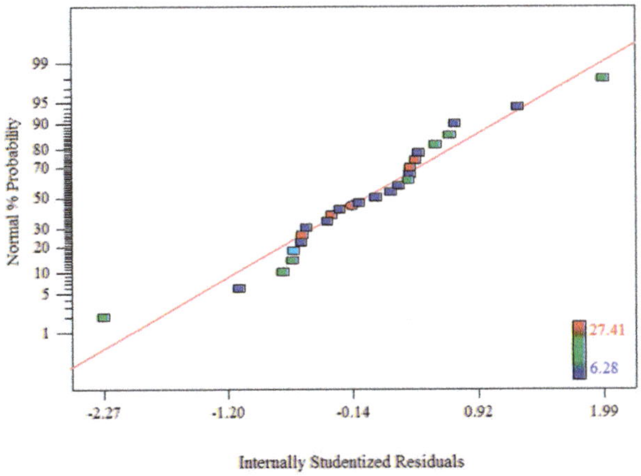

Figure 1. Normal probability plot of the response errors for the barium sulfate mass model.

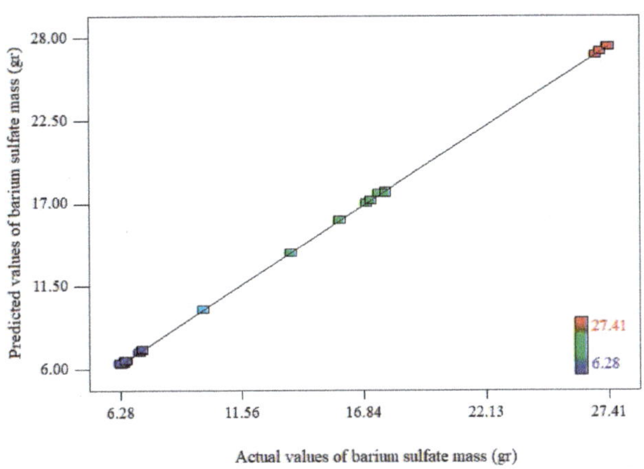

Figure 2. The difference between the actual and predicted values of the mass of barium sulfate (in grams).

Figure 3 illustrates that with the increase in the concentrations of barium chloride and sulfuric acid, the mass of the produced barium sulfate rises correspondingly. However, an excessive concentration of sulfuric acid in the reaction environment inhibits the growth of barium sulfate crystals, leading to a reduction in the overall material production. The feed rate also impacts the mass of the produced material; specifically, an increase in the feed rate results in a lower mass of produced material. Temperature does not significantly influence this process.

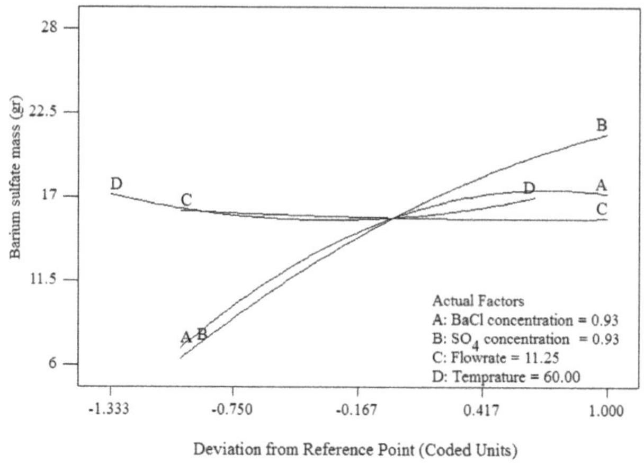

Figure 3. Effect of parameters on the mass of produced barium sulfate. A: barium chloride concentration; B: sulfate concentration; C: feed rate; D: temperature.

Figure 4 showcases the influence of critical parameter interactions on the mass of produced barium sulfate. The response surface plots offer an in-depth visualization of how varying conditions impact the synthesis process. In Figure 4a, the effects of barium chloride and sulfate concentrations are analyzed. The data clearly indicate that simultaneously increasing the concentrations of these reactants significantly enhances the mass of barium sulfate produced, underscoring that higher reactant concentrations promote more efficient barium sulfate precipitation. Figure 4b illustrates the relationship between $BaCl_2$ concentration and temperature. The findings reveal that while an increase in $BaCl_2$ concentration markedly boosts production, the temperature, ranging from 20 °C to 80 °C, exerts a minimal influence on the mass. This observation suggests that within the examined range, temperature is not as crucial a parameter as reactant concentration. Figure 4c investigates the impact of sulfate concentration and flow rate. It becomes evident that reducing the flow rate leads to a notable increase in the mass of barium sulfate produced, even with slight variations in sulfate concentration. This indicates that lower flow rates extend the residence time, thereby enhancing precipitation efficiency. Figure 4c suggests that reducing the flow rate leads to increased mass production, likely because lower flow rates provide more time for the barium sulfate particles to nucleate and grow. This is particularly important in continuous flow reactors, where optimizing the residence time can lead to higher yields without the need for excessive increases in reactant concentrations. Figure 4d examines the interaction between temperature and flow rate. The analysis reveals that variations in flow rate have a more substantial impact on the mass of barium sulfate at lower temperatures. As temperatures rise, this effect lessens, highlighting a complex interaction where specific temperature ranges dampen the influence of flow rate on the precipitation process. In conclusion, the response surface methodology illustrated in Figure 4 emphasizes the critical role of optimizing reactant concentrations and flow rate to maximize the efficiency of barium sulfate production, while temperature assumes a more secondary role.

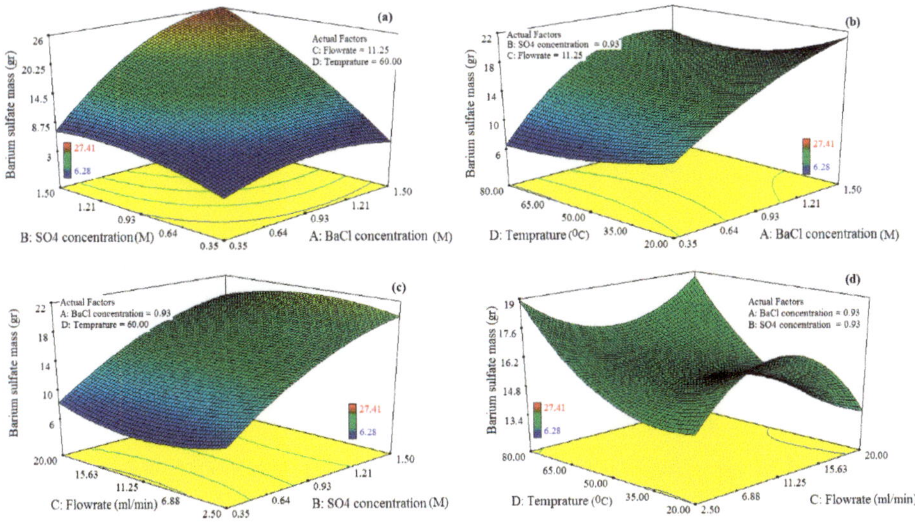

Figure 4. Impact of different parameter interactions on the mass of produced barium sulfate. (**a**) Barium chloride and sulfate concentration; (**b**) barium chloride concentration and temperature; (**c**) feed rate and sulfate concentration; (**d**) feed rate and temperature.

3.1.2. Density of Barium Sulfate

Figures 5 and 6 present, respectively, the normal probability plot and the distribution of actual response values versus predicted values for the density model. The data in both of these plots align with the model's predictions, further validating its accuracy and consistency.

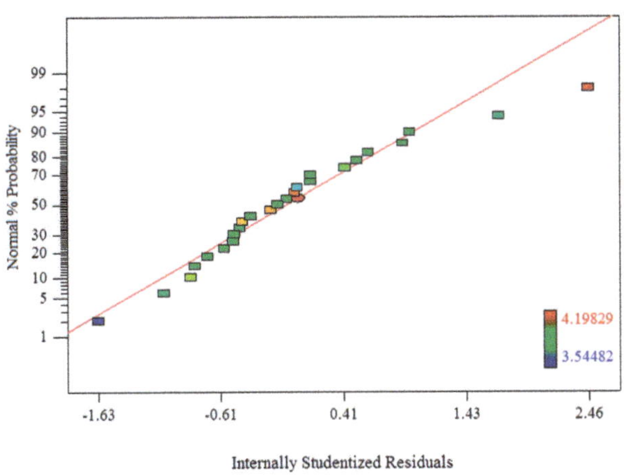

Figure 5. Normal probability plot of the response errors for the barium sulfate density model.

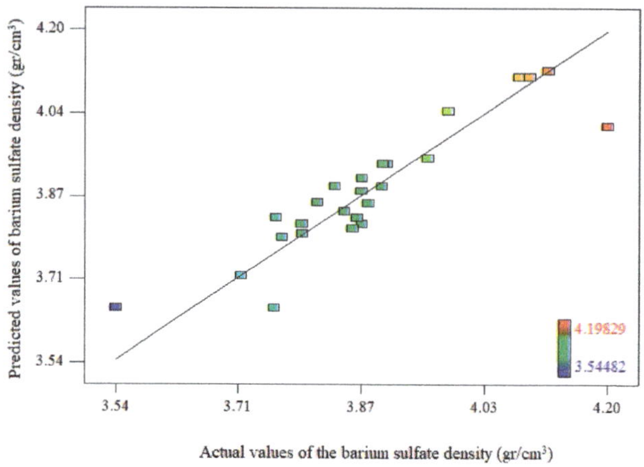

Figure 6. The difference between the actual and predicted values of the density of barium sulfate.

Additionally, Figure 7 demonstrates that as the concentration of barium chloride increases, the density of barium sulfate rises. Conversely, increasing the concentration of sulfate and temperature results in a decrease in density. The feed rate does not have a significant effect on the density of the barium sulfate produced using sulfuric acid.

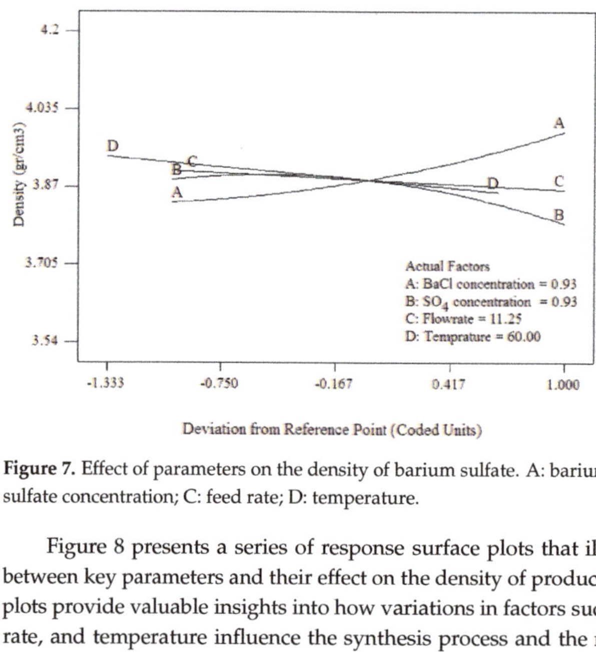

Figure 7. Effect of parameters on the density of barium sulfate. A: barium chloride concentration; B: sulfate concentration; C: feed rate; D: temperature.

Figure 8 presents a series of response surface plots that illustrate the interactions between key parameters and their effect on the density of produced barium sulfate. These plots provide valuable insights into how variations in factors such as concentration, flow rate, and temperature influence the synthesis process and the resulting barium sulfate density. The slight increase in density with increasing concentrations of barium chloride and sulfate (as shown in Figure 8a) reflects the tendency for denser products to form under higher concentrations of reactants. This may be due to enhanced crystal growth and aggregation of particles, leading to a more compact structure. In industrial applications, higher density barium sulfate can be beneficial for certain applications, such as in drilling fluids, where product density is a critical parameter. Figure 8b explores the relationship between sulfate concentration and flow rate. The findings indicate that changes in flow rate have a relatively minor effect on the density of barium sulfate, particularly when the sulfate concentration is held constant. This suggests that adjustments in flow rate alone may not significantly impact the overall density unless they are coupled with changes in other factors. It reinforces the idea that under certain conditions, flow rate alone does not play a dominant role in influencing the product's characteristics. In Figure 8c, the influence of temperature, in conjunction with sulfate concentration, on barium sulfate density is assessed. The data show that within the temperature range of 70 °C to 80 °C, fluctuations in temperature have a negligible effect on density. This suggests that when sulfate concentration is controlled, temperature changes do not substantially affect the precipitation process, pointing to the limited role of temperature within the studied range. Finally, Figure 8d examines the interaction between temperature and flow rate. The relatively flat surface in this plot indicates that neither temperature nor flow rate, within the ranges tested, significantly influences the density of barium sulfate. This suggests that there exists a threshold beyond which further changes in these parameters do not substantially alter the final density of the product. This observation implies that optimizing these variables might be less critical compared to other factors, such as concentration. The relatively insensitive nature of barium sulfate density to variations in temperature and flow rate, as indicated in Figure 8c,d, suggests that temperature and flow rate adjustments beyond certain thresholds do not significantly alter the physical characteristics of the product. This finding is consistent with the concept of thermodynamic equilibrium, where the synthesis conditions may reach a point beyond which further changes in temperature or flow rate yield diminishing returns in terms of product properties. Overall, the response surface methodology presented in Figure 8 emphasizes the importance of optimizing reactant concentrations to achieve the desired barium sulfate density. It also illustrates the relatively minor impact of temperature and flow rate variations within the experimental ranges. These insights are valuable for the development of efficient and cost-effective strategies for barium sulfate synthesis, offering a pathway for optimizing the process and improving product yield.

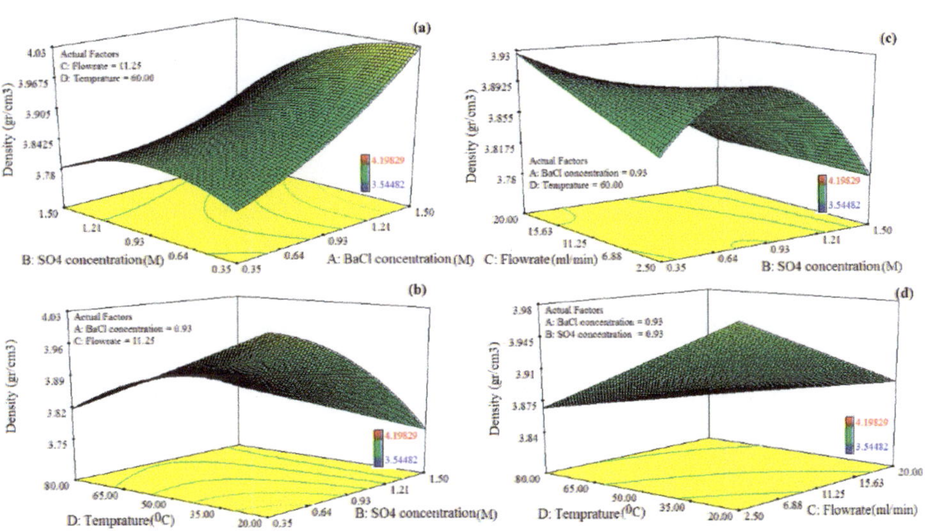

Figure 8. Impact of parameter interactions on the density of barium sulfate. (**a**) Barium chloride and sulfate concentration; (**b**) sulfate concentration and temperature; (**c**) feed rate and sulfate concentration; (**d**) feed rate and temperature.

3.2. Optimization Conditions

The optimal conditions for this process were determined by the software as follows: under the experimental optimum conditions, $BaCl_2$ concentration = 1.5 M, SO_4^{2-} concentration = 0.41 M, flow rate = 2.53 mL/min, and temperature = 20.03 °C. Under these conditions, the software predicted the production of 9.21 g of barium sulfate with a density of 4.25 g/cm^3. The optimal condition tests were repeated three times, yielding the results presented in Table 6. This shows that the experimental values closely align with the predicted output, confirming the reliability of the model.

Table 6. Experimental results and average values for barium sulfate synthesis under optimized conditions.

Test No.	Mass (g)	Density (g/cm^3)
1	9.11	4.29
2	9.27	4.2
3	9.23	4.26
Average	9.20	4.25

Table 7 displays the results from the X-ray fluorescence (XRF) analysis of barium sulfate synthesized under optimum conditions. The XRF analysis clearly supports the notion that the synthesis method employed produces barium sulfate of exceptional purity (high elemental purity), which is critical for industries that require high-quality materials, such as pharmaceuticals, ceramics, and cosmetics. The minute quantities of impurities (e.g., Al_2O_3 and SrO) are within acceptable thresholds for proposed applications, ensuring the integrity of the end product for stringent industrial uses. These minor quantities of impurities are due to the use of industrial-grade purity reagents. The most notable observation from the data is the absence of significant impurities such as silica and potassium oxide, which are often problematic in synthesizing materials. The presence of minor traces of aluminum oxide (Al_2O_3) and strontium oxide (SrO) further emphasizes the controlled nature of the synthesis process, where such components are kept to minimal levels. This result affirms the robustness of the synthesis method, which could be of great interest to industries requiring barium sulfate with high performance characteristics. The average purity of the barium sulfate was determined to be 96.08%.

Table 7. The XRF analysis results of synthesized barium sulfate.

Component	%	Component	%
Potassium Oxide (K_2O)	0.04	Cadmium Oxide (CdO)	0.025
Sodium Oxide (Na_2O)	0.12	Silver Oxide (Ag_2O)	0.023
Iron(III)7Oxide (Fe_2O_3)	0.011	Copper Oxide (CuO)	0.014
Strontium Oxide (SrO)	0.33	Chlorine (Cl)	0.93
Aluminum Oxide (Al_2O_3)	0.82	Calcium Oxide (CaO)	0.083
Barium Oxide (BaO)	62.44	Phosphorus Pentoxide (P_2O_5)	0.017
Sulfur Trioxide (SO_3)	33.64	Magnesium Oxide (MgO)	0.027

The high purity of the synthesized barium sulfate, coupled with its low levels of contaminants, demonstrates the suitability of the product for critical industrial applications where impurity levels must be carefully controlled. The method presented here could serve as a valuable contribution to improving existing production systems for barium sulfate, especially for industries that demand high grades of material for their processes.

Furthermore, the scalability of this synthesis method holds promise for industrial-scale production. Given the efficiency in achieving high purity and the low operational costs, this technique could offer a pathway for more sustainable and economically viable large-scale manufacturing of barium sulfate.

4. Conclusions

This study successfully developed a methodology for the synthesis of barium sulfate at the laboratory scale. Using the D-optimal design, critical reaction parameters—such as barium chloride and sulfate concentrations—were systematically optimized. Under the experimental optimum conditions—$BaCl_2$ concentration = 1.5 M, SO_4^{2-} concentration = 0.41 M, flow rate = 2.53 mL/min, and temperature= 20.03 °C—9.21 g barium sulfate with a density of 4.25 g/cm^3 can be achieved. The experimental results and corresponding average values for barium sulfate synthesis under optimized conditions demonstrate that, on average, 9.20 g of barium sulfate with a density of 4.25 g/cm^3 was produced. This shows that the experimental values closely align with the predicted output, confirming the reliability of the model. The analysis of barium sulfate produced in the optimum conditions confirmed the purity, with a barium oxide content of 62.44% and sulfur trioxide at 33.64%, verifying the material's industrial applicability. The robustness and reliability of the optimization process were validated through rigorous statistical analysis, which demonstrated strong predictive capabilities with high F-values and R^2 indicators for both mass and density outcomes. Future research should focus on scaling the process for large-scale industrial applications and exploring additional variables that could further enhance production efficiency and sustainability.

Author Contributions: Conceptualization, M.S. (Mehdi Safari), F.S.H., A.S. and B.R.; Methodology, A.S., F.S.H. and M.S. (Mehdi Safari); Software, A.S. and M.S. (Mohammad Sadatipour); Validation, A.S., F.S.H. and M.S. (Mehdi Safari); Formal analysis, A.S., F.S.H. and M.S. (Mehdi Safari); Investigation, A.S., F.S.H. and M.S. (Mehdi Safari); Resources, M.S. (Mehdi Safari); Data curation, A.S.; Writing—original draft, F.S.H., M.S. (Mohammad Sadatipour) and M.S. (Mehdi Safari); Writing—review & editing, F.S.H. and M.S. (Mehdi Safari); Visualization, M.S. (Mohammad Sadatipour) and F.S.H.; Supervision, F.S.H., B.R. and M.S. (Mehdi Safari); Project administration, F.S.H. and M.S. (Mehdi Safari); Funding acquisition, M.S. (Mehdi Safari). All authors have read and agreed to the published version of the manuscript.

Funding: This research received no external funding.

Data Availability Statement: The data presented in this study are available on request from the corresponding author.

Conflicts of Interest: The authors declare no conflict of interest.

References

Dye, W.M.; Mullen, G.A.; Gusler, W.J. Field-Proven Technology To Manage Dynamic Barite Sag. In Proceedings of the IADC/SPE Drilling Conference, Miami, FL, USA, 21–23 February 2006. [CrossRef]

Al Jaberi, J.; Bageri, B.; Elkatatny, S.; Patil, S. Primary Investigation of Barite-Weighted Water-Based Drilling Fluid Properties. *ACS Omega* **2023**, *8*, 2155–2163. [CrossRef]

Deville, J.P. Drilling fluids. In *Fluid Chemistry, Drilling and Completion*; Elsevier: Amsterdam, The Netherlands, 2022; pp. 115–185.

Bruton, J.R.; Bacho, J.; Newcaster, J. The Future of Drilling-Grade Barite Weight Material—A Case for a Substitute Specification. In Proceedings of the SPE Annual Technical Conference and Exhibition, San Antonio, TX, USA, 24–27 September 2006; pp. SPE-103135-MS.

Afolayan, D.O. Assessment of Beneficiation Process of a Nigerian Barite Source for Utilization as Oil Drilling Mud. Ph.D. Thesis, African University of Science and Technology, Abuja, Nigeria, 2017.

6. Matos, C.T.; Devauze, C.; Planchon, M.; Ewers, B.; Auberger, A.; Dittrich, M.; Wittmer, D.; Latunussa, C.; Eynard, U.; Mathieux *Material System Analysis of Nine Raw Materials: Barytes, Bismuth, Hafnium, Helium, Natural Rubber, Phosphorus, Scandium, Tantalu and Vanadium*; Publications Office of the European Union Luxembourg: Luxembourg, 2021.
7. Sun, S.; Ding, H.; Zhou, H. Preparation of TiO_2-coated barite composite pigments by the hydrophobic aggregation method ar their structure and properties. *Sci. Rep.* **2017**, *7*, 10083. [CrossRef]
8. Bahadori, A.; Zahedi, G.; Zendehboudi, S. Estimation of potential barium sulfate (barite) precipitation in oilfield brines using simple predictive tool. *Environ. Prog. Sustain. Energy* **2012**, *32*, 860–865. [CrossRef]
9. Al-Rubaii, M.; Al-Shargabi, M.; Al-Shehri, D. A Novel Model for the Real-Time Evaluation of Hole-Cleaning Conditions wi Case Studies. *Energies* **2023**, *16*, 4934. [CrossRef]
10. Aremu, M.; Arinkoola, A.; Salam, K.; Ogunmola, E. Potential of local pH control additives for corrosion inhibition in water ba drilling fluids. *Pet. Coal* **2017**, *59*, 611–619.
11. Elkatatny, S. Mitigation of barite sagging during the drilling of high-pressure high-temperature wells using an invert emulsic drilling fluid. *Powder Technol.* **2019**, *352*, 325–330. [CrossRef]
12. Maile, F.J. Colorants in coatings. *Phys. Sci. Rev.* **2021**, *6*, 707–789. [CrossRef]
13. Gysau, D. *Fillers for Paints*; Vincentz Network: Hannover, Germany, 2017.
14. Abel, A. Pigments for paint. In *Paint and Surface Coatings: Theory and Practice*; Woodhead Publishing Cambridge: Cambridge, U 1999; pp. 91–165.
15. Emblem, H.G.; Hargreaves, K. Preparation, Properties and Uses of Barium Compounds. *Rev. Inorg. Chem.* **1995**, *15*, 109–1 [CrossRef]
16. Qi, L.; Cölfen, H.; Antonietti, M. Control of Barite Morphology by Double-Hydrophilic Block Copolymers. *Chem. Mater.* **2000**, 2392–2403. [CrossRef]
17. Carretero, M.I.; Pozo, M. Clay and non-clay minerals in the pharmaceutical industry: Part I. Excipients and medical applicatior *Appl. Clay Sci.* **2009**, *46*, 73–80. [CrossRef]
18. El-Mezayen, A. Mineralogical and chemical studies on some minerals used in pharmaceutical industries in Egypt. *Al-Azhar Pharm. Sci.* **2014**, *50*, 181–190. [CrossRef]
19. Safari, M.; Harris, M.; Deglon, D. The effect of energy input on the flotation kinetics of galena in an oscillating grid flotation ce In Proceedings of the 27th International Mineral Processing Congress (IMPC 2014), Santiago, Chile, 20–24 October 2014.
20. Lopresti, M.; Palin, L.; Alberto, G.; Cantamessa, S.; Milanesio, M. Epoxy resins composites for X-ray shielding materia additivated by coated barium sulfate with improved dispersibility. *Mater. Today Commun.* **2021**, *26*, 101888. [CrossRef]
21. Krejci, M.R.; Wasserman, B.; Finney, L.; McNulty, I.; Legnini, D.; Vogt, S.; Joester, D. Selectivity in biomineralization of barium an strontium. *J. Struct. Biol.* **2011**, *176*, 192–202. [CrossRef]
22. Zhang, L.; Yang, Y.; Zhang, J.; Zhou, X.; Dong, H.; Zhou, Y. Barium sulfate aspiration: Severe chemical pneumonia induced by massive reflux of contrast medium during small bowel barium enema. *Forensic. Sci. Int.* **2015**, *253*, e16–e19. [CrossRef] [PubMed
23. Kleinsasser, L.J.; Warshaw, H. Perforation of the sigmoid colon during barium enema: Report of a case with review of th literature, and experimental study of the effect of barium sulfate injected intraperitoneally. *Ann. Surg.* **1952**, *135*, 560. [CrossRe
24. Oskarsson, A. Barium. In *Handbook on the Toxicology of Metals*; Elsevier: Amsterdam, The Netherlands, 2022; pp. 91–100.
25. Aziz, H.A.; Ghazali, M.F.; Hung, Y.-T.; Wang, L.K. Toxicity, source, and control of barium in the environment. In *Handbook Advanced Industrial and Hazardous Wastes Management*; CRC Press: Boca Raton, FL, USA, 2017; pp. 463–482.
26. Matta, G.; Gjyli, L.; Kumar, M.A. *Inorganic Toxicity: Environment & Human Health*; LAP LAMBERT Academic Publishin Saarbrücken, Germany, 2018.
27. Vanhoof, C.; Bacon, J.R.; Ellis, A.T.; Fittschen, U.E.A.; Vincze, L. 2019 atomic spectrometry update—A review of advances in X-ra fluorescence spectrometry and its special applications. *J. Anal. At. Spectrom.* **2019**, *34*, 1750–1767. [CrossRef]
28. Janssens, K. X-ray based methods of analysis. In *Comprehensive Analytical Chemistry*; Elsevier: Amsterdam, The Netherlands, 200 Volume 42, pp. 129–226.
29. Crabb, E.; Moore, E. *Metals and Life Chapter 9*; RSC Publishing: Cambridge, UK, 2010.
30. Reznik, R.; Troy, H. *Source Assessment: Major Barium Chemicals*; Environmental Protection Agency, Office of Research an Development, Industrial Environmental Research Laboratory: Research Triangle Park, NC, USA, 1978.
31. Zhang, W.; Zhou, Y.; Zhu, J.; Pan, Y. New Clean process for barium sulfide preparation by barite reduction with elemental sulfu *Ind. Eng. Chem. Res.* **2014**, *53*, 5646–5651. [CrossRef]
32. Safari, M.; Hoseinian, F.S.; Deglon, D.; Filho, K.L.L.; Souza, T. Investigation of the reverse flotation of hematite in three differen types of laboratory flotation cells. In Proceedings of the 29th International Mineral Processing Congress (IMPC 2018), Moscov Russia, 17–21 September 2018; pp. 1376–1383.
33. Dibello, P.M.; Manganaro, J.L.; Aguinaldo, E.R.; Mahmood, T.; Lindahl, C.B. *Barium Compounds*; Kirk-Othmer Encyclopedia o Chemical Technology: Hoboken, NJ, USA, 2000.

Dorozhkin, S.V. Calcium orthophosphates and human beings: A historical perspective from the 1770s until 1940. *Biomatter* **2012**, *2*, 53–70. [CrossRef]

Stark, A.I. *Barium Sulfate Precipitation Kinetics: An Experimental and Analytical Comparison with and Without Scale Inhibitors*; The University of Manchester (United Kingdom): Manchester, UK, 2005.

Jaffe, E.B. *Abstracts of the Literature on Synthesis of Apatites and Some Related Phosphates, Part I and Part II*; US Department of the Interior, Geological Survey: Reston, VA, USA, 1950.

Toner, R.K. *The Determination of the Factors Affecting the Commercial Production of Barium-Chloride From Barite and Calcium-Chloride*; Purdue University: West Lafayette, IN, USA, 1939.

Shreve, R.N.; Toner, R.K. Barium Chloride from Barite and Calcium Chloride Factors Affecting Production. *Ind. Eng. Chem.* **1940**, *32*, 568–573. [CrossRef]

Rizkalla, E.N. Kinetics of the crystallisation of barium sulphate. Effect of additives, stirring rate and barium: Sulphate ratio on the rate of precipitation. *J. Chem. Soc. Faraday Trans. 1 Phys. Chem. Condens. Phases* **1983**, *79*, 1857–1867. [CrossRef]

Adhikari, G.; Sudborough, J.; Watson, H. The production of barium salts from barytes. *J. Indian Inst. Sci.* **1922**, *5*, 83–105.

Abdulla, S.A.E. *Barium Sulphate: Problems, Solutions and Applications*; University of Surrey (United Kingdom): Surrey, UK, 2010.

Tasbolatovna, Y.S.; Ilyich, K.A. The most important technologies of receiving barium sulfate from barite. *Eur. Sci.* **2020**, *6*, 30–34.

Smith, T.A.; Chen, D. Green Synthesis Approaches in Industrial Mineral Processing. *Miner. Eng.* **2021**, *169*, 106956. [CrossRef]

Hoseinian, F.S.; Safari, M.; Deglon, D. Ion flotation kinetic predictions using empirical and phenomenological models. *Miner. Eng.* **2024**, *210*, 108645. [CrossRef]

Baloyi, N.P.; Nheta, W.; Sibanda, V.; Safari, M. Mineralogical Insights into PGM Recovery from Middle Group (1–4) Chromite Tailings. *Minerals* **2024**, *14*, 924. [CrossRef]

Hoseinian, F.S.; Irannajad, M.; Safari, M. Effective factors and kinetics study of zinc ion removal from synthetic wastewater by ion flotation. *Sep. Sci. Technol.* **2016**, *52*, 892–902. [CrossRef]

Lee, R.M.; Hassan, M. Barium Sulfate in Modern Radiographic Imaging: Purity Requirements and Synthesis Challenges. *Mater. Chem. Phys.* **2023**, *302*, 127680. [CrossRef]

Taguta, J.; Safari, M.; Govender, V.; Chetty, D. Investigating the Amenability of a PGM-Bearing Ore to Coarse Particle Flotation. *Minerals* **2023**, *13*, 698. [CrossRef]

Nazari, S.; Hoseinian, F.S.; Li, J.; Safari, M.; Khoshdast, H.; Li, J.; He, Y. Synergistic effect of grinding time and submicron (nano) bubbles on the zeta potential state of spent lithium-ion batteries: A gene expression programming approach. *J. Energy Storage* **2023**, *70*, 107942. [CrossRef]

Hoseinian, F.S.; Rezai, B.; Safari, M. A new kinetic model for zinc ion removal from synthetic wastewater. In Proceedings of the 10th European Metallurgical Conference (EMC 2019), Düsseldorf, Germany, 23–26 June 2019; pp. 1409–1416.

Hasanizadeh, I.; Khoshdast, H.; Safari, M.; Asgari, K.; Rahmanian, A. Influence of Carbonated Pyrolysis Oil Recycled from Scrap Tires on Metallurgical Efficiency of Coal Flotation. *Minerals* **2024**, *14*, 765. [CrossRef]

ASTM D2330-02(2017); Standard Test Method for Density, Relative Density (Specific Gravity), or API Gravity of Light Hydrocarbons by Pressure Hydrometer. ASTM International: West Conshohocken, PA, USA, 2017.

Safari, M.; Harris, M.; Deglon, D.; Filho, L.L.; Testa, F. The effect of energy input on flotation kinetics. *Int. J. Miner. Process.* **2016**, *156*, 108–115. [CrossRef]

Hassanzadeh, A.; Safari, M.; Khoshdast, H.; Güner, M.K.; Hoang, D.H.; Sambrook, T.; Kowalczuk, P.B. Introducing key advantages of intensified flotation cells over conventionally used mechanical and column cells. *Physicochem. Probl. Miner. Process.* **2022**, *58*, 155101. [CrossRef]

Carelse, C.; Manuel, M.; Chetty, D.; Taguta, J.; Safari, M.; Youlton, K. The flotation behaviour of liberated Platinum Group minerals in Platreef ore under reduced reagent conditions. *Miner. Eng.* **2022**, *190*, 107913. [CrossRef]

Wang, X.; Bu, X.; Ni, C.; Zhou, S.; Yang, X.; Zhang, J.; Alheshibri, M.; Peng, Y.; Xie, G. Effect of scrubbing medium's particle size on scrubbing flotation performance and mineralogical characteristics of microcrystalline graphite. *Miner. Eng.* **2021**, *163*, 106766. [CrossRef]

Montgomery, D.C. *Design and Analysis of Experiments*, 9th ed.; John Wiley & Sons: Hoboken, NJ, USA, 2017.

Khormali, A.; Ahmadi, S. Prediction of barium sulfate precipitation in dynamic tube blocking tests and its inhibition for waterflooding application using response surface methodology. *J. Pet. Explor. Prod. Technol.* **2023**, *13*, 2267–2281. [CrossRef]

Li, D.; Li, M.; Pan, J.; Yan, Z. The deposition kinetics of barium sulfate scale: Model development. *Front. Mater.* **2023**, *10*, 1198176.

60. Piotrowski, K.; Wojtas, M.; Berenjian, A. Influence of Hydrodynamic Conditions on Precipitation Kinetics of Barium Sulfa Experimental and Modeling Approach. *Processes* **2022**, *10*, 146.
61. U.S. Geological Survey. *Mineral Commodity Summaries 2021*; U.S. Geological Survey: Reston, VA, USA, 2021; 200p. [CrossRef]

minerals

MDPI

ticle

ffects of Grinding Parameters on Galena Particle Size)istribution and Flotation Performance

engchi Guo [1], Yuankun Yang [2], Shengli Yu [2], Yanming Wu [1,2,*], Guohua Gu [1,*], Yanhong Wang [1], Qingke Li [1] and ınyu Chen [1]

1 School of Minerals Processing and Bioengineering, Central South University, Changsha 410083, China; 225612105@csu.edu.cn (M.G.); yanhong.wang@csu.edu.cn (Y.W.); liqingke2022@163.com (Q.L.); 225611034@csu.edu.cn (J.C.)

2 Zijin Mining Group Co., Ltd., Longyan 364200, China; yang.yuankun@zijinmining.rs (Y.Y.); yu.shengli@zijinmining.rs (S.Y.)

* Correspondence: wu_yanming@zijinmining.com (Y.W.); guguohua@csu.edu.cn (G.G.)

Abstract: The processing of low-grade, lead-containing practical ores requires fine grinding to liberate galena and enhance flotation recovery. The ball mill is still one of the most common approaches used in industry for fine grinding. This study investigated the effect of the grinding parameters in a ball mill on the fine grinding product of galena and on flotation performance. The grinding product had a particle size below 30 μm, which was classified into +25 μm, −25 + 10 μm, and −10 μm fractions. Grinding experiments showed that modifications to the grinding concentration, media proportion, and filling ratio exerted significant effects on the yields of the +25 μm and −10 μm fractions. Flotation experiments showed that the yield of −10 μm particles negatively affected the flotation performance of galena. Discrete element method simulation results revealed that an increase in the motion velocity of the media group enhanced attrition effects during fine grinding, promoting the generation of −10 μm particles. The higher yield of −10 μm particles facilitated a smaller contact angle and smaller agglomerate size, resulting in lower recovery. To optimize the particle size distribution and improve fine-grained galena flotation recovery, it is essential to reduce the attrition of the grinding media on the mineral.

Keywords: grinding parameters; fine grinding; galena; particle size distribution; fine particle flotation

1. Introduction

Lead is recognized as a critical national strategic resource, playing an indispensable role in the electrical, mechanical, military, and metallurgical industries, and serves as a cornerstone of national economic development [1–3]. Galena, the most widely distributed lead-bearing mineral, is the primary raw material for lead metal production. As global demand for lead continues to rise, the recycling and utilization of microfine-grained galena resources have garnered increasing attention, with corresponding research efforts intensifying [4]. Fine grinding is a well-established technique for enhancing the liberation of microfine mineral particles and optimizing flotation recovery efficiency [5]. Given the growing emphasis on resource efficiency and technological innovation, investigating the impact of fine grinding on the flotation performance of microfine-grained galena is of significant scientific and practical importance.

Ball mills and stirred mills are commonly used equipment for fine ore grinding [5]. Although stirred mills have advantages, such as lower energy consumption and higher grinding efficiency compared to ball mills [6], the design and development of large-scale stirred mills within a short timeframe remain challenging due to structural limitations and material strength constraints [7]. Consequently, traditional ball mills remain the most widely used fine grinding equipment, due to their high throughput and simple structure [8]. Given their widespread application, investigating the effects of fine grinding in ball mill systems remains highly significant.

The surface properties of galena particles in the ball mill grinding product vary with particle size. Generally, the finer the galena particles, the more complex their surface properties [9–11]. It has been proven that under identical flotation reagent conditions, −25 μm galena particles are more soluble and oxidize more readily than +25 μm particles, enhancing their surface hydrophilicity [12], and that galena particles smaller than 10 μm are more affected by iron oxidation species from grinding media than those in the intermediate size range of −53 + 10 μm [13]. Therefore, in the fine grinding stage of galena, achieving an optimal particle size distribution is crucial for ensuring effective separation and enhancing flotation recovery [14,15], and it is essential to study the factors influencing the particle size distribution of galena grinding products. It has been established that the material composition [16,17], shape [18,19], and size [20] of the grinding media, along with variations in the mill mechanisms [21–23], directly impact the particle size distribution of fine-ground mineral products. Additionally, determining appropriate grinding parameters, such as the grinding concentration, media proportion, and filling ratio, is also critical for achieving the desired particle size distribution in fine-ground products, thereby ensuring optimal separation efficiency [24,25]. When adjusting the grinding parameters, the variation regularity of the particle size distribution of galena fine grinding products remains unclear and requires further investigation.

During the fine grinding stage, mineral particles are generally ground to below 30 μm [26]. This paper investigates galena with 65% of particles finer than 74 μm, which is ground to achieve a particle size below 30 μm in a ball mill system. The study examines the effects of the grinding parameters, including grinding concentration, media proportion, and filling ratio, on the particle size distribution of the fine-ground galena product. The variation in flotation recovery confirms the reliability of the analysis results for the changes in grinding product particle size distribution. The research findings provide new insights into how the grinding parameters affect the particle size distribution of galena during fine grinding, thus facilitating the recycling of fine-grained galena resources.

2. Materials and Methods

2.1. Pure Minerals

The galena sample used in this study was purchased from the Mingfa Mineral Specimen Company in Guangzhou, China and is sourced from a medium-low temperature hydrothermal deposit. The X-ray fluorescence (XRF) analysis, which was performed using an Axios mAX device made by PANalytical, which was located in Almelo, The Netherlands, showed that it contained 85.83% Pb, which meant that the purity of galena was 99.09%. Following hand selection, crushing, and grinding, the galena sample exhibited the particle size distribution shown in Figure 1, with 65% of the particles smaller than 74 μm. As shown in Figure 1, since the minerals were crushed below 200 μm in the crushing stage, and grinding in the grinding stage targeted minerals under 100 μm from the crushing product, the minerals in the 100–200 μm range were not ground. Thus, the laser particle size analysis results of galena showed a second peak, with the maximum particle size around 200 μm.

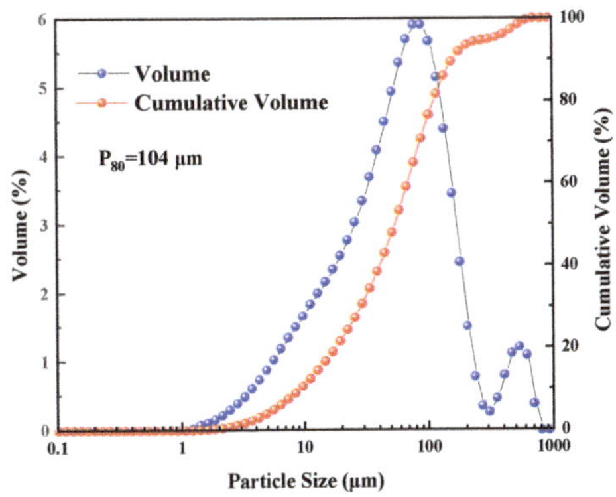

Figure 1. The laser particle size analysis results of galena (the feed material for the grinding experiments).

2.2. Grinding Experiments

The experimental process is shown in Figure 2. Grinding experiments were conducted using a stainless-steel cylindrical mill with a grinding chamber size of 80 mm (D) × 100 mm (L). Literature reports have demonstrated that small-scale ball mill jars have been effectively utilized in laboratory settings to investigate the impact of grinding conditions on the properties of ground products and have consistently shown stable outcomes and well-defined trends [27]. For each test, 60 g of galena sample, deionized water and stainless-steel balls were placed in the ball mill, which had an effective volume of 500 mL, and ground for 10 min at a mill speed of 109 r/min; the critical speed of the mill is 149 r/min. The grinding parameters were set as detailed in Table 1. When 65% of the particles in the feed material are smaller than 74 μm, relevant research has demonstrated that in laboratory-scale tests, the fine grinding efficiency of minerals is optimal when the media sizes are selected as 5 mm, 3 mm, and 2 mm [28]. The grinding concentration is defined as the ratio of the mass of solids to the mass of pulp, which is equivalent to the ratio of the mass of solids to the sum of the mass of solids and the mass of water. The media proportion refers to the combination ratio of the grinding media of different sizes or types within a ball mill. The filling ratio is the percentage of the grinding media volume relative to the mill's effective volume.

Table 1. The grinding parameters and their values examined in the experiments.

Title	Value		
Grinding concentration	50%	67%	75%
Media proportion	5 mm:3 mm:2 mm = 4:3:3	5 mm:3 mm:2 mm = 4:4:2	5 mm:3 mm:2 mm = 5:3:2
Filling ratio	30%	40%	50%

Pre-tests were carried out to explore the grinding parameters appropriate to grind 60 g of galena samples to below 30 μm under the above-mentioned conditions of a mill speed of 109 r/min and a grinding time of 10 min. The results show that a grinding concentration of 67%, a media ratio of 4:4:2, and a filling ratio of 40% are required to achieve a grinding product size of less than 30 μm. Based on these conditions, the grinding concentration, media proportion, and filling ratio were successively varied to obtain the fine-ground products for each set of grinding parameters, which were then used for subsequent experiments and analysis. For example, when adjusting the grinding concentration, the media ratio was 4:4:2 and the filling ratio was 40%. When adjusting the media ratio, the grinding concentration was 67% and the filling ratio was 40%. When adjusting the filling ratio, the grinding concentration was 67% and the media ratio was 4:4:2.

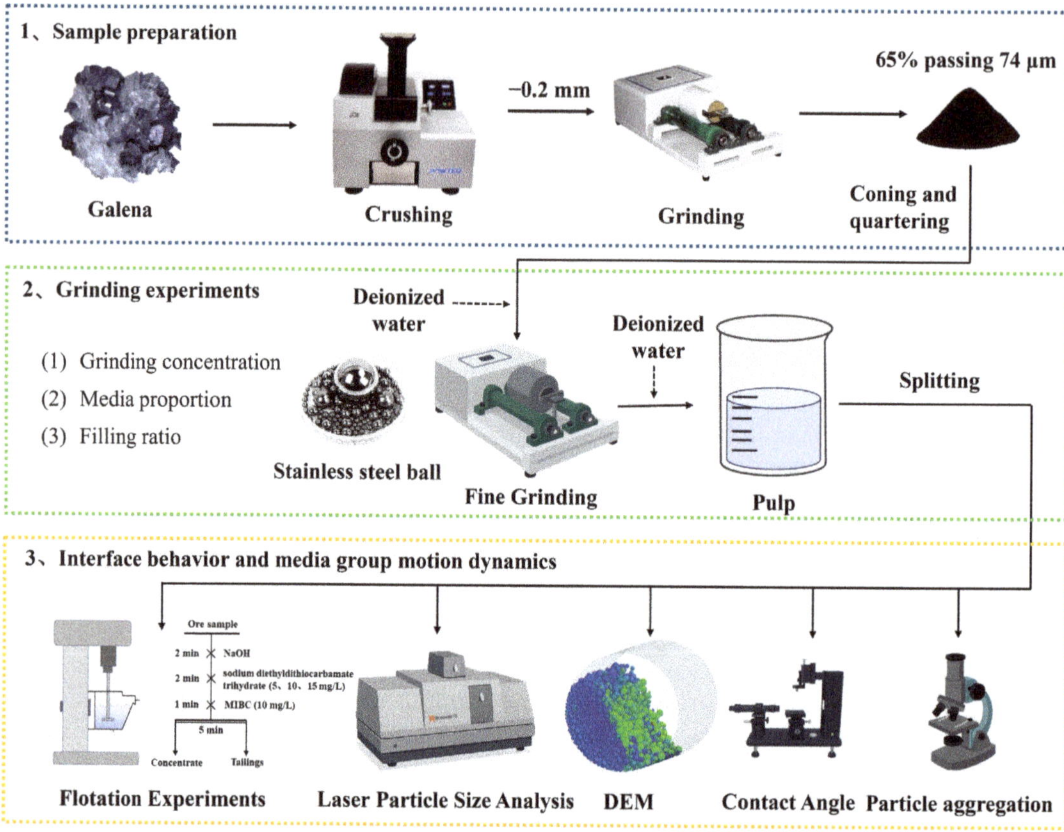

Figure 2. Schematic diagram of the experimental process.

2.3. Laser Particle Size Analysis

The particle size distribution of the fine-ground products was analyzed using a Mastersizer 2000 laser particle size analyzer manufactured by Malvern Instruments, located in Malvern, UK. During the experiments, 1 g of the sample was dispersed in 900 mL of water in a beaker. Dispersion was facilitated by a stirrer operating at 2500 rpm, and ultrasonic treatment was applied for 1 min. The average particle size was automatically calculated by the analyzer. In the laser particle size analysis experiments, multiple measurements were taken for the same sample. The average particle size and the yields of the three average particle size fractions were obtained after processing the particle size distribution data.

2.4. Flotation Experiments

In the flotation experiments, analytical grade NaOH, industrial-grade sodium diethyldithiocarbamate trihydrate, and analytical-grade methyl isobutyl carbinol (MIBC) were used as the pH adjusters, collectors, and frothers, respectively.

For each flotation experiment, 15 g of the fine-ground galena product was placed in a 140 mL flotation tank, and deionized water was added; the pulp density used during flotation was 15.79 wt.%. The mixture was stirred at 1992 rpm for 2 min. The slurry pH was adjusted to 9, followed by the sequential addition of the collector sodium diethyldithiocarbamate trihydrate at dosages of 5, 10, 15 g/Mg with two minutes of stirring and the frother MIBC at a dosage of 10 g/Mg with one minute of stirring. Flotation was conducted for 5 min. After flotation, the products were filtered, dried, weighed, and the galena recovery was calculated based on the weights of the concentrate and tailing products. The specific calculation formula is expressed as follows:

$$\varepsilon = \frac{m_1}{m_1+m_2} \times 100\% \tag{1}$$

In the formula, ε denotes the galena recovery, m_1 is the concentrate product weight, and m_2 is the tailing product weight.

2.5. Discrete Element Simulation

In this study, the grinding process in a ball mill was numerically simulated using the discrete element method (DEM) to elucidate the motion behavior of the grinding media and investigate the mechanisms by which the grinding parameters influence the fine-grinding efficiency of galena. The ball mill model and the reference model of the steel ball media used in the DEM simulation are shown in Figure 3. Figure 3 only shows the model of the 5 cm ball; models of the 3 cm and 2 cm balls were also used in the simulation, matching the actual grinding experiment. The Hertz–Mindlin (no slip) contact model was employed in the simulation, as it effectively predicts the motion state of the media group within the ball mill [29]. The particle contact parameters are listed in Table 2. Each simulation ran for a total duration of 5 s, divided into 50 time steps, with a step size of 0.1 s. This study utilized EDEM commercial software (EDEM 2022, Altair, Troy, MI, USA).

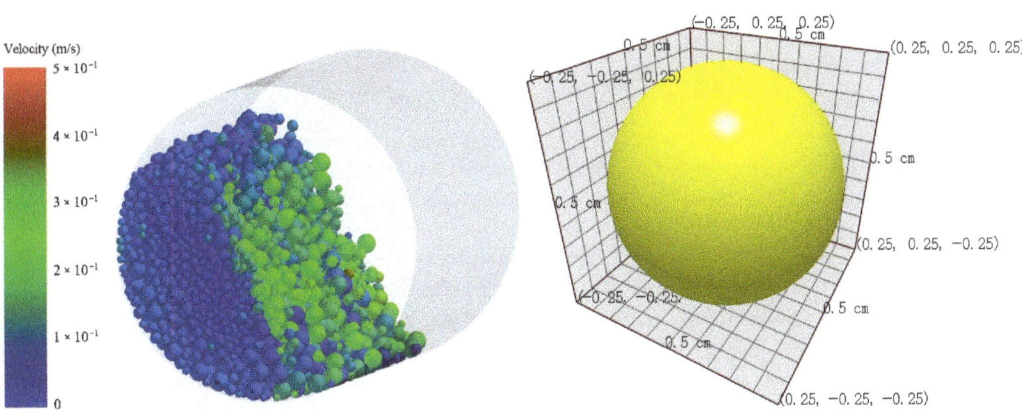

Figure 3. The model used in the DEM simulations.

Table 2. Particle contact parameters (derived from the manufacturer's test report).

Type of Contact	Steel-Steel	Steel Ball-Cylinder
Rolling friction coefficient	0.3	0.2
Static friction coefficient	0.5	0.5
Restitution coefficient	0.15	0.15

2.6. Contact Angle Measurements

After taking a sample identical to the feed for the flotation test from the finely ground product, it was mixed with deionized water and stirred in a beaker under the flotation reagent system with a collector dosage of 10 g/Mg at 1992 r/min, and when no collector was added, no treatment was performed. After stirring, the pulp was filtered and dried, and the dried product was then pressed into a smooth slice. A JC2000DM contact angle meter was used to measure the contact angle of the samples through the sessile drop method. In this method, a stable water droplet with a diameter of 3–4 mm was precisely deposited onto the slice surface using a microsyringe for contact angle measurement. Each sample was measured three times at different locations, and the average value was recorded.

2.7. Observation of Hydrophobic Agglomerates

An optical microscope (OLYMPUS CX33, PLENIC-Pro, Hangzhou, China) was used to observe the aggregate particle size of the fine-ground galena products. After fine grinding, a 1 g sample was taken from the product and stirred with deionized water in a beaker under the flotation reagent system with a collector dosage of 10 g/Mg at 1992 r/min, and when no collector was added, no treatment was performed. After stirring, 1 mL of the pulp was extracted using a pipette and placed on a glass slide to ensure a uniform monolayer distribution of the aggregates. The slide was then placed under the optical microscope, and the sample was examined at 100× magnification. Subsequently, the average agglomerate size was analyzed using OPLENIC-Pro software (V1.95). In each stirring test, ten samples were observed for analysis and discussion to ensure data accuracy. The calculation of the average agglomerate size and error was based on the measurements of the largest agglomerate size during each of the ten observations. The photos that best represent the average agglomerate size were selected for presentation [30].

3. Results and Discussion

3.1. Particle Size Distribution of the Fine-Ground Galena Products

Figures 4 and 5 present the particle size distribution curve and the average particle size of the fine-ground galena products under different grinding concentrations, media proportions, and filling ratios. As shown in Figure 5, when the grinding concentration increased from 50% to 75% or the media proportion shifted from 4:3:3 to 5:3:2, the average particle size of the fine-ground product initially decreased and then increased. This can be attributed to the fact that lower grinding concentrations and specific media proportions provide optimal grinding conditions. Low grinding concentrations reduce the buffer effect between grinding media, enhancing the impact and attrition forces on the galena particles. However, excessively high concentrations may lead to overloading, reducing the efficiency of energy transfer to the particles [24]. Similarly, the optimal media proportion (4:3:3) balances the impact and grinding actions of different media sizes. Deviating from this proportion disrupts the grinding dynamics, resulting in coarser particles at higher concentrations and specific media ratios. In contrast, increasing the filling ratio from 30% to 50% caused a gradual reduction in average particle size. A higher filling ratio increases the number of grinding media in the mill, leading to more frequent collisions and grinding actions on the galena particles. This results in more effective size reduction [31]. The shifts in the particle size distribution curves under various grinding process parameters shown in Figure 4 demonstrate well the reliability of the analysis results regarding changes in average particle size.

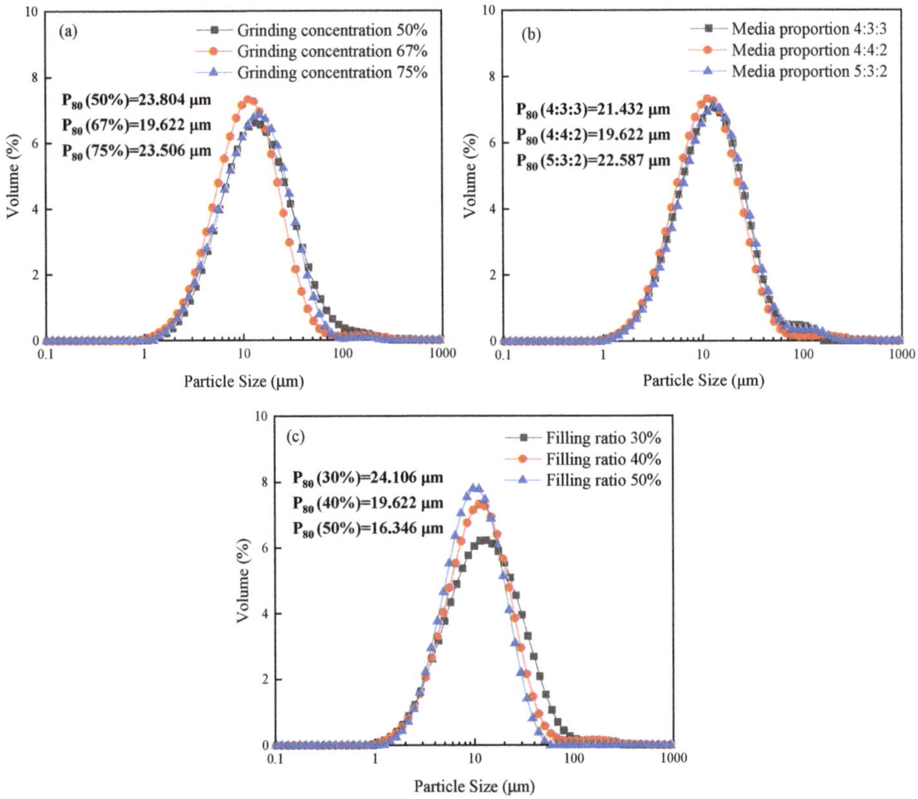

Figure 4. Particle size distribution curve of the fine-ground galena products obtained from different (**a**) grinding concentration, (**b**) media proportion, and (**c**) filling ratio.

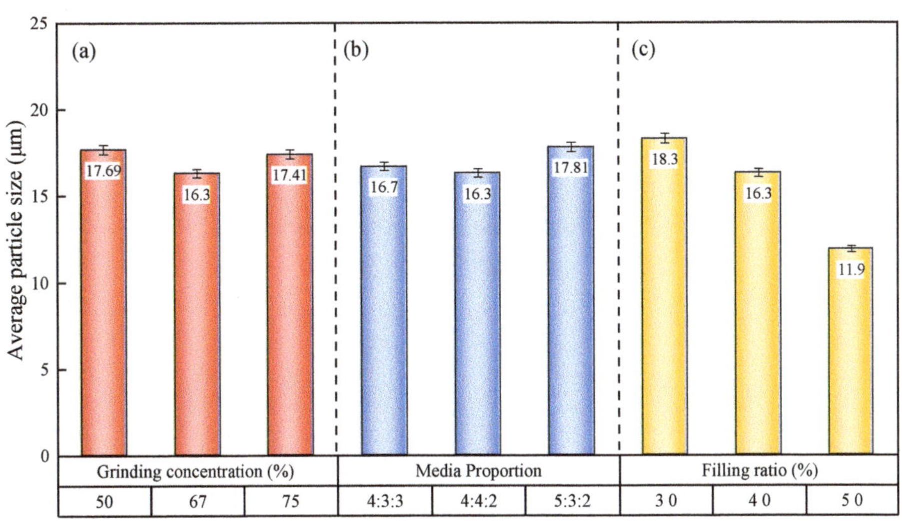

Figure 5. Average particle size of the fine-ground galena products obtained from different (**a**) grinding concentration, (**b**) media proportion, and (**c**) filling ratio.

The influence of the grinding parameters on the particle size distribution of the fine-ground products was further explored. The fine-ground products with a particle size below 30 µm obtained under different grinding parameters were classified into +25 µm, −25 + 10 µm, and −10 µm fractions, and the yield of each size fraction was calculated, as shown in Figure 6. Figure 6 illustrates that as the grinding concentration increased from 50% to 75%, the yield of +25 µm particles initially decreased and then increased. This trend is consistent with the average particle size variation, indicating that moderate grinding intensities favor the reduction of coarse particles. The yield of −25 + 10 µm particles remained relatively stable, suggesting that this size fraction is less sensitive to changes in grinding concentration within the tested range. Meanwhile, the yield of −10 µm particles first increased and then decreased, indicating that both excessively high and low grinding concentrations are not conducive to the formation of fine particles. This is similar to the trend observed when the media proportion was adjusted from 4:3:3 to 5:3:2. This suggests that both the grinding concentration and media proportion significantly affect the grinding efficiency and particle size distribution [32]. When the filling ratio increased from 30% to 50%, the yield of +25 µm particles gradually decreased, the yield of −25 + 10 µm particles showed negligible variation, while the yield of −10 µm particles progressively increased. This indicates that higher filling ratios enhance grinding efficiency.

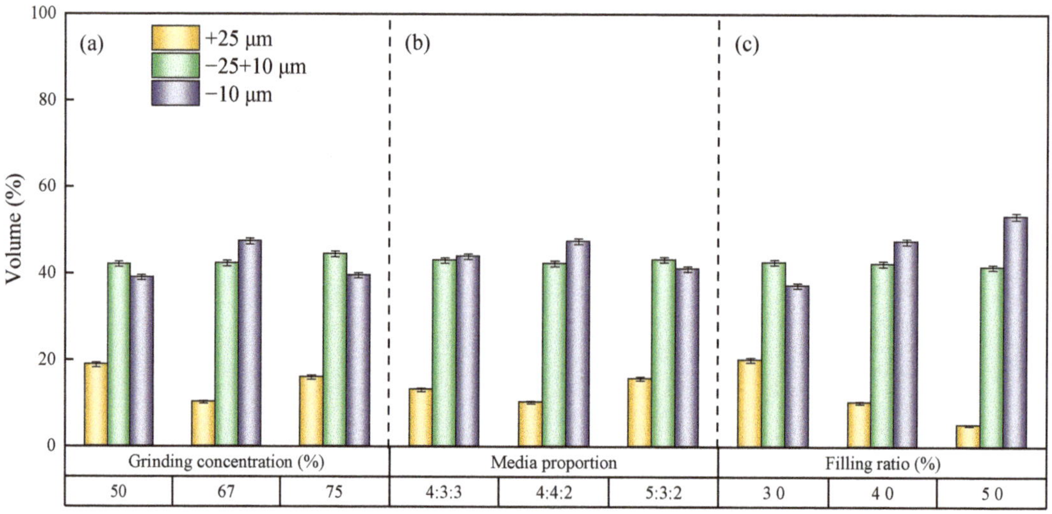

Figure 6. Yield of each size fraction in the fine-ground galena products obtained from different (**a**) grinding concentration, (**b**) media proportion, and (**c**) filling ratio.

3.2. Flotation Experiments on the Fine-Ground Galena Products

The recovery of the fine-ground galena products under different grinding parameters is shown in Figure 7. As Figure 7 shows, under a constant collector dosage, increasing grinding concentration from 50% to 75% initially decreased the galena recovery, followed by a subsequent increase, consistent with the variation observed when the media proportion was adjusted from 4:3:3 to 5:3:2. When the filling ratio increased from 30% to 50%, the galena recovery gradually declined. Combined with the particle size distribution analysis results, the variation in +25 µm particles, particularly the yield of −10 µm particles, significantly impacted galena recovery. Specifically, a higher yield of −10 µm particles led to lower galena recovery. When the filling rate reached 50%, the yield of −10 µm particles in the fine-ground product was highest at 53.38%, while the galena recovery was lowest. This indicates that the flotation performance is closely related to the particle size distribution of the ground products. Coarser particles (+25 μm) may not be sufficiently liberated, leading to lower flotation recovery. Finer particles (−10 μm) may increase the risk of over-pulverization, which can negatively affect flotation efficiency. The −25 + 10 μm fraction represents a balance between liberation and over-pulverization, making it the most suitable size range for optimal flotation performance. Therefore, understanding and controlling the particle size distribution through grinding parameters is crucial for optimizing the flotation process and achieving higher galena recovery and concentrate quality.

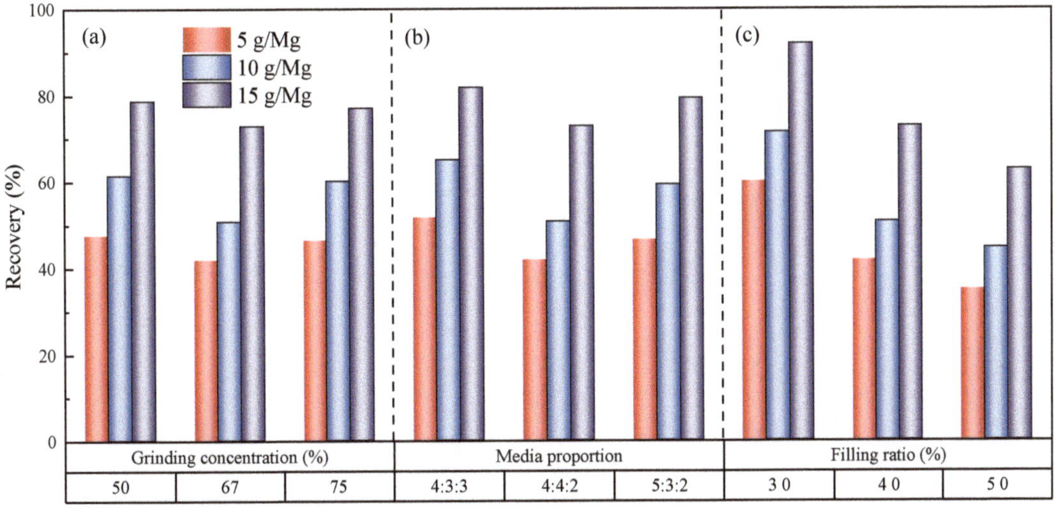

Figure 7. The effect of different (**a**) grinding concentration, (**b**) media proportion, and (**c**) filling ratio on the recovery of the fine-ground galena products (the collector dosages were 5, 10, and 15 g/Mg).

3.3. Results of DEM

The motion state of the grinding media in the mill was analyzed using discrete element simulation to elucidate the influence of different grinding parameters on fine grinding performance in the grinding experiments [33,34]. The velocity cloud diagram of the media group is presented in Figure 8. Generally, a higher motion velocity of the media group corresponds to a larger effective grinding area, enhancing material breakage [35]. As shown in Figure 8, with increasing grinding concentration or adjusting the media proportion from 4:3:3 to 5:3:2, the motion velocity of the media group first increased and then decreased. In contrast, the motion velocity of the media group increased progressively with an increase in the filling ratio. At a filling ratio of 50%, the media group exhibited the highest motion velocity, resulting in the most effective crushing and grinding of materials. This phenomenon occurs because an increase in the filling rate leads to a greater overall media volume, raising the shoulder position of the media balls within the rotating cylinder. Consequently, the potential energy increases and is progressively converted into kinetic energy during descent, thereby increasing the motion velocity of the media balls [36].

Under different grinding parameters, the normal and tangential forces of media collisions in the mill are shown in Figure 9. Generally, the normal force primarily facilitates impact breakage of the material, with the intensity of media impact determined by its magnitude. The tangential force primarily contributes to attrition. Among them, the grinding action primarily affects the yield of −10 μm particles in the grinding product, showing a positive correlation [37–39].

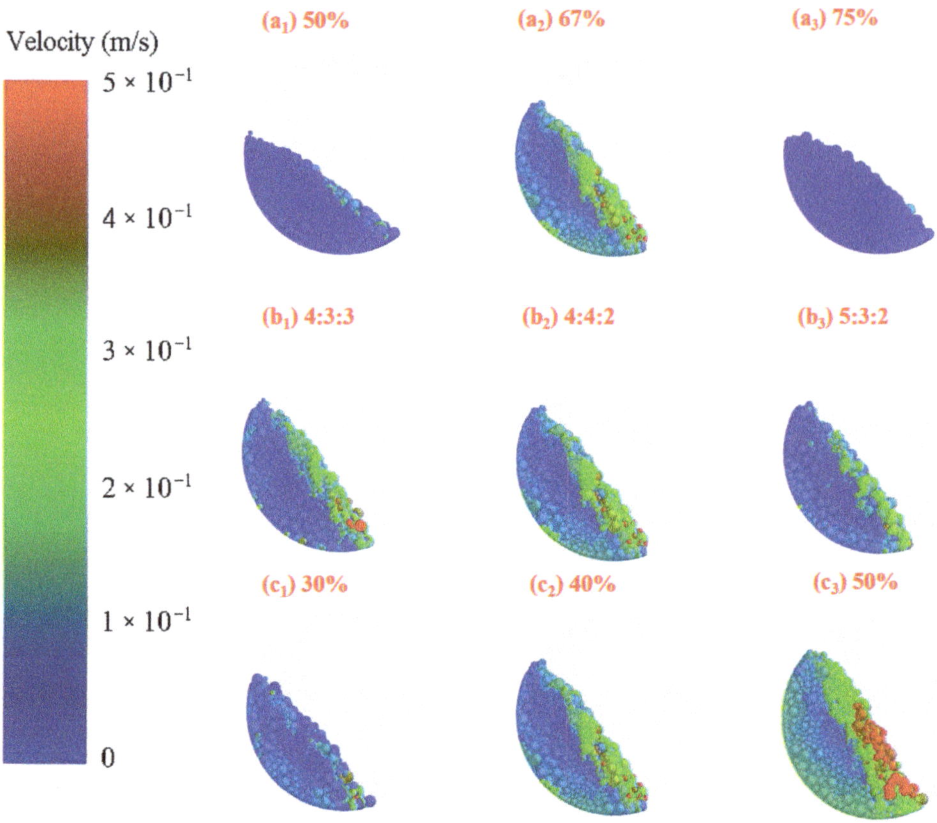

Figure 8. Velocity cloud of media group under different (**a**) grinding concentration, (**b**) media proportion, and (**c**) filling ratio.

In Figure 9, adjustments to the grinding parameters altered the normal and tangential forces exerted by the grinding media on the minerals, thereby influencing both impact and attrition effects, which, in turn, affected the particle size distribution of the fine-ground products. It should be noted that during fine grinding in a ball mill, attrition is the primary mechanism. As the particle size of the mineral decreases, the grinding action exerted by the media on the mineral becomes stronger, while the impact action gradually weakens. However, the simulation results indicate that the normal force, which characterizes the impact action, is relatively large. This may be due to the higher mill speed, which causes some grinding media balls to be in a state of free fall. Additionally, since mineral particles were not included in the simulation and the effective contact area between media balls is small, grinding action is difficult to generate. Therefore, in the simulation results, the normal force representing the impact action is greater than the tangential force [36,40]. Nevertheless, when keeping other simulation conditions constant, the influence of the grinding process parameters on the tangential force, which characterizes the grinding action, can still be observed. Therefore, it can still be concluded that adjusting the grinding process parameters changes the grinding action, thereby affecting the particle size distribution of the finely ground product. The normal and tangential forces exerted by the media on the mineral exhibited a unimodal trend with increasing grinding concentration or when the media ratio was adjusted from 4:3:3 to 5:3:2; in contrast, both forces demonstrated a monotonic increase with rising filling rate. At a filling ratio of 50%, the normal and tangential forces reached their maximum values, resulting in the strongest impact and attrition effects exerted by the media on the minerals. This observation is consistent with Figures 4–6, where the fine-ground galena products exhibited the smallest average particle size and the highest yield of −10 µm particles.

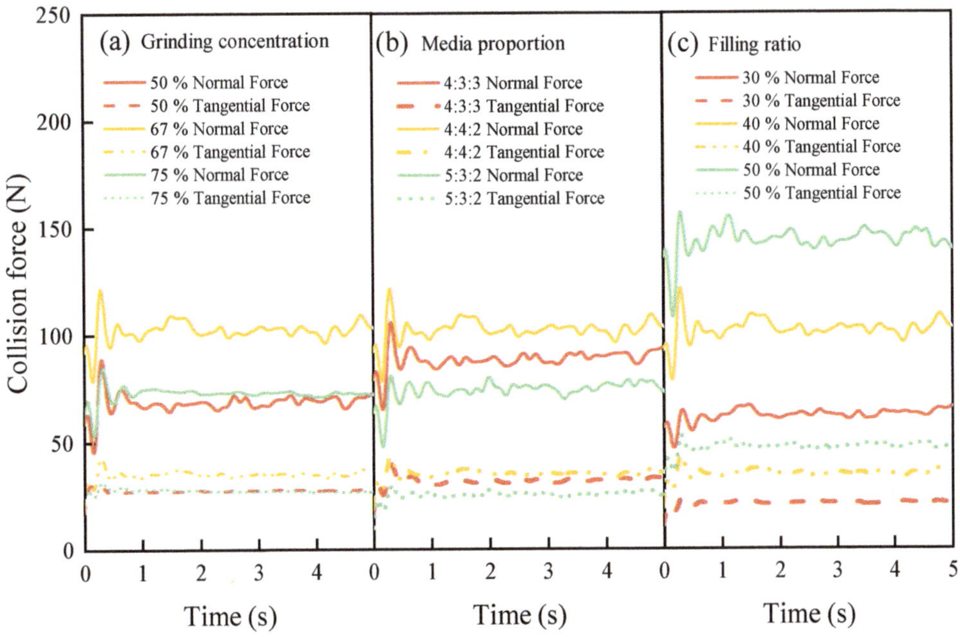

Figure 9. The variation in normal and tangential collision forces with time of different (**a**) grinding concentration, (**b**) media proportion, and (**c**) filling ratio.

3.4. Contact Angle Analysis

The surface wettability of minerals varies with particle size [41–43], while grinding parameters influence the particle size distribution of the fine-ground galena product. As shown in Figures 10 and 11, the contact angle values presented in the bar chart on the right, along with the error bars, are derived from three measurements taken at different locations on the sample. Each photo on the left is an exemplary photo of the three photos for each sample. The effects of the grinding parameters on the surface contact angle of galena are presented under two conditions: without a collector and with a collector dosage of 10 g/Mg. Based on the particle size distribution analysis results, Figures 10 and 11, when the grinding parameters were adjusted, an increase in the yield of the −10 µm fraction in fine-ground galena products correlated with a smaller surface contact angle and lower galena recovery. Specifically, at a filling rate of 30%, the yield of the −10 µm fraction was minimized, while the galena surface contact angle was maximized, resulting in the highest galena recovery.

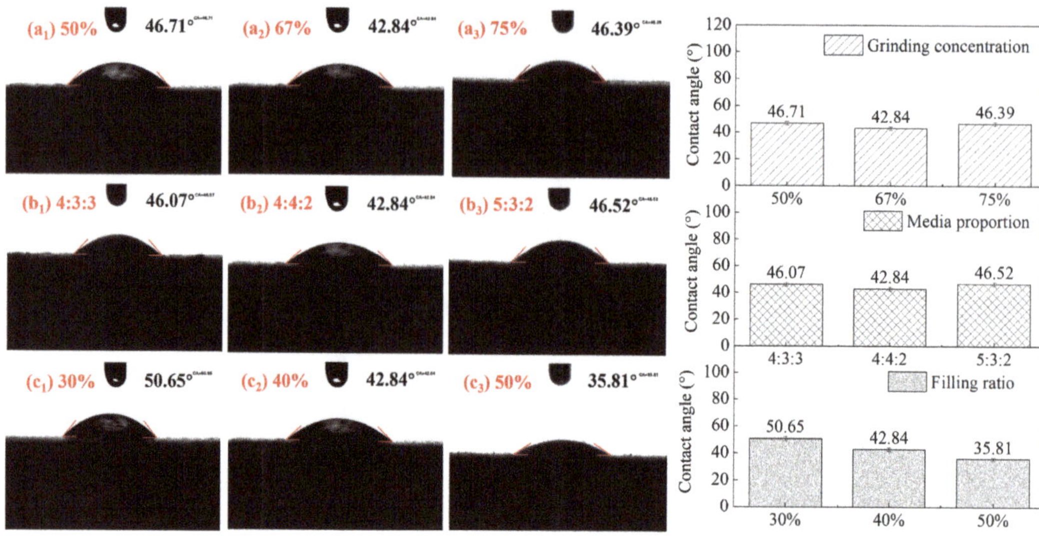

Figure 10. Influence of (**a**) grinding concentration, (**b**) media proportion, and (**c**) filling rate on the surface contact angle of the fine-ground galena products (pH = 9, sodium diethyldithiocarbamate trihydrate = 0).

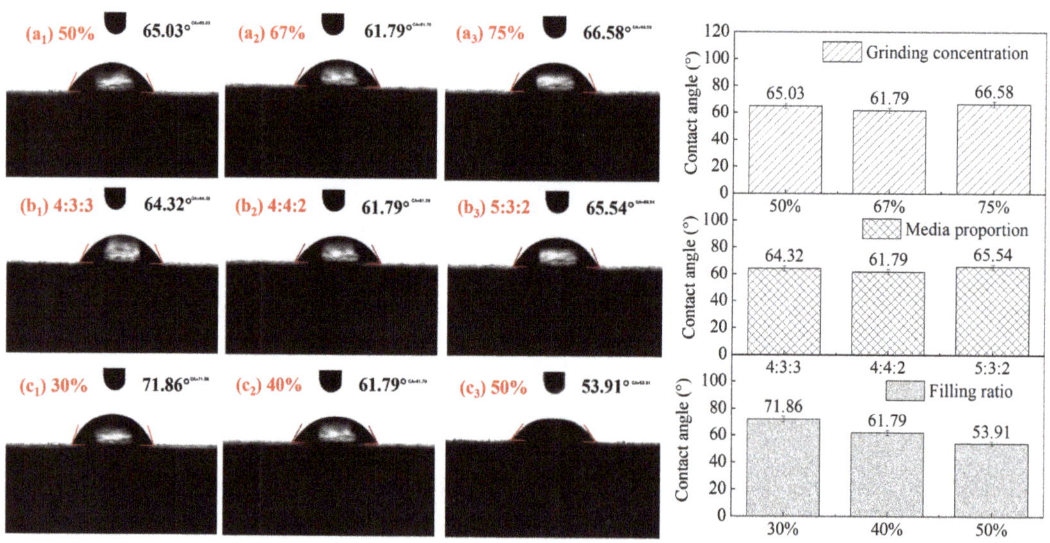

Figure 11. Influence of (**a**) grinding concentration, (**b**) media proportion, and (**c**) filling rate on the surface contact angle of the fine-ground galena products (pH = 9, sodium diethyldithiocarbamate trihydrate = 10 g/Mg).

3.5. Hydrophobic Agglomeration

Due to variations in the surface wettability of minerals with different particle sizes, the agglomeration behavior of particles changes during flotation, with larger agglomerate size exhibiting better floatability [44,45]. The effects of the grinding parameters on the average agglomerate size of the fine-ground galena products, under conditions without a collector and with a collector dosage of 10 g/Mg, are shown in Figures 12 and 13. The average agglomerate size values and error bars presented in the bar chart on the right are derived from the measurement of the largest agglomerate size during each of the ten observations. Each photo on the left is selected from ten observations of a single stirring experiment to best represent the average agglomerate size and the most representative agglomerates. Based on the particle size analysis, Figures 12 and 13, when grinding parameters such as the grinding concentration, media proportion, and filling ratio were adjusted, the −10 μm yield in fine-ground galena products rose, the average agglomerate size decreased, and the galena recovery declined. Specifically, at a filling rate of 50%, the −10 μm yield peaked, the average agglomerate size was minimal, and the galena recovery was lowest.

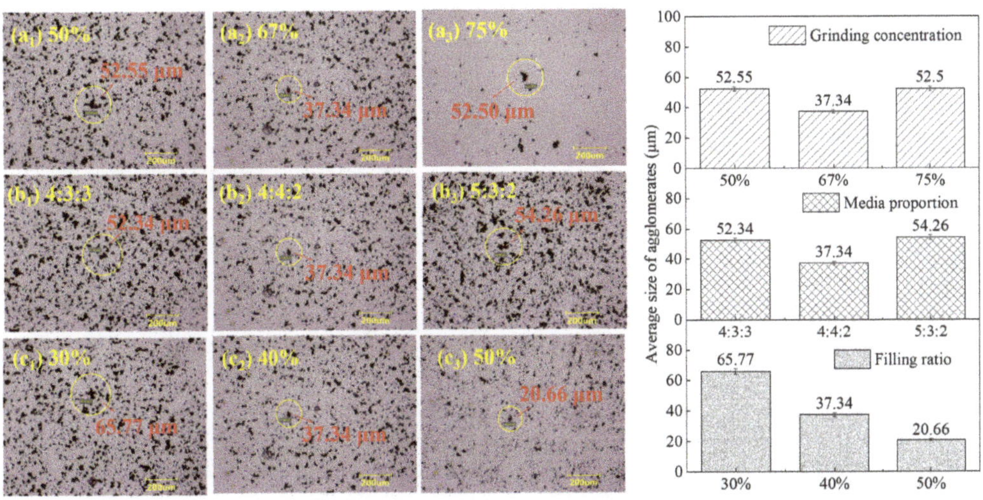

Figure 12. Influence of (**a**) grinding concentration, (**b**) media proportion, and (**c**) filling rate on the average agglomerate size of the fine-ground galena products (pH = 9, sodium diethyldithiocarbamate trihydrate = 0).

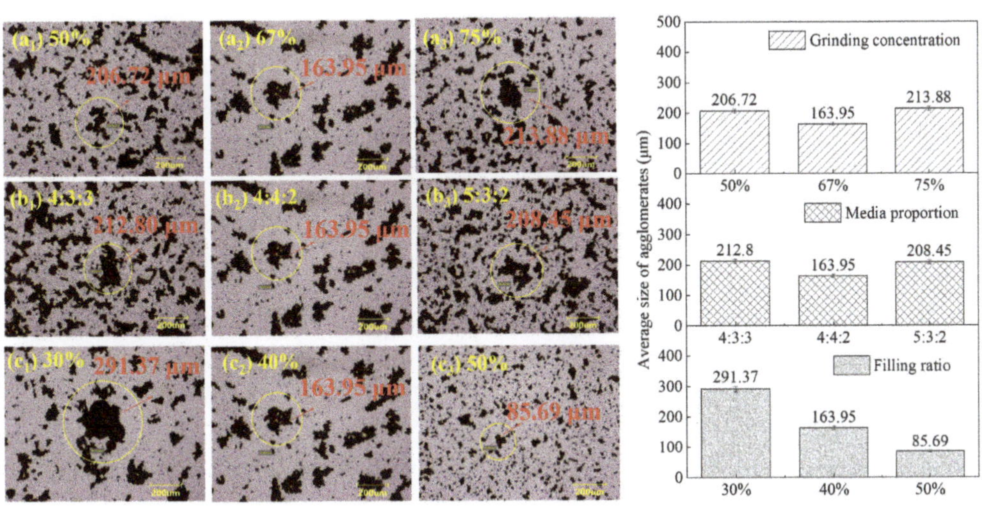

Figure 13. Influence of (**a**) grinding concentration, (**b**) media proportion, and (**c**) filling rate on the average agglomerate size of the fine-ground galena products (pH = 9, sodium diethyldithiocarbamate trihydrate = 10 g/Mg).

3.6. Effects of Grinding Parameters on Particle Size Distribution and Recovery of the Fine-Ground Galena Products

As can be seen from Figure 14, changes in the grinding parameters led to different tangential forces exerted by the media on minerals, influencing the generation of −10 μm particles (Figure 9), leading to differences in the average particle size, +25 μm yield, and particularly, the −10 μm yield of the fine-ground galena products (Figure 6), which ultimately influenced interfacial behaviors, such as the surface contact angle (Figures 10 and 11), agglomeration (Figures 12 and 13), and flotation (Figure 7). As shown in Figure 15, the yield of the −10 μm particle size fraction in the finely ground galena product shows a positive linear correlation with the tangential force, which characterizes the attrition, and a negative linear correlation with the contact angle of the mineral surface, the average aggregate size, and the galena recovery.

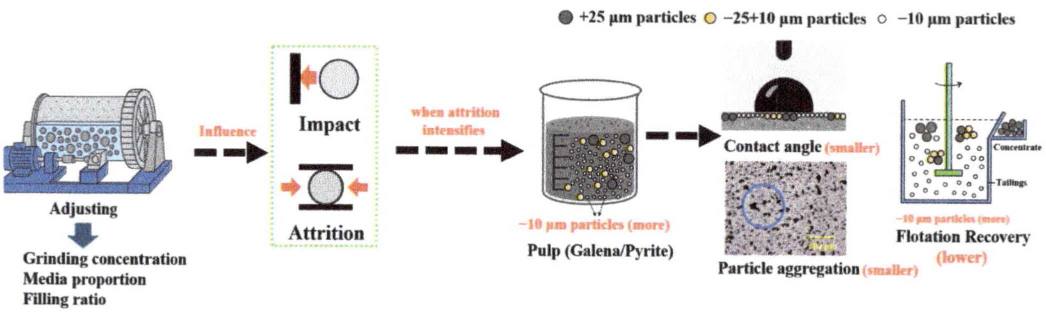

Figure 14. Effects of grinding parameters on particle size distribution and recovery of the fine-ground galena products.

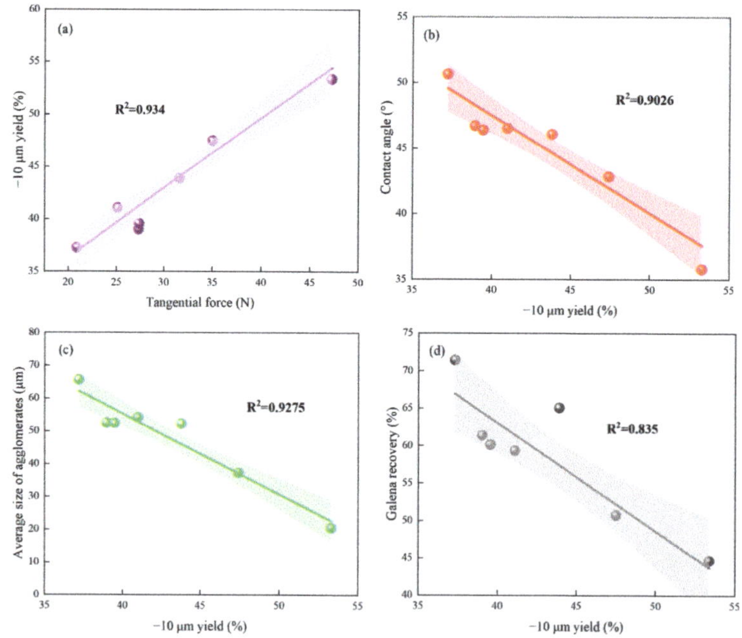

Figure 15. Linear fitting results: **(a)** −10 μm yield vs. tangential force (plot in purple); **(b)** Contact angle and average aggregate size vs. −10 μm yield (plot in red); **(c)** Average aggregate size vs. −10 μm yield (plot in green); **(d)** Galena recovery vs. −10 μm yield (plot in gray).

4. Conclusions

The fine-ground product with a particle size below 30 μm was classified into three size fractions: +25 μm, −25 + 10 μm, and −10 μm. Changes in the grinding concentration, media proportion, and filling ratio significantly affected the yields of the +25 μm and −10 μm fractions, and the yield of the −25 + 10 μm particle size fraction remained relatively unchanged. When the filling rate increased from 30% to 50%, the yield of the +25 μm particle size fraction decreased by 15.04%, and the yield of the −10 μm particle size fraction increased by 16.10%. Discrete element simulation results revealed that increasing the motion velocity of the media group and enhancing attrition effects by adjusting the grinding parameters during fine grinding promoted the generation of −10 μm particles.

The −10 μm particle yield significantly affected mineral flotation behavior, with higher yields leading to lower galena recovery. Surface wettability and agglomeration analyses showed that the higher yield of −10 μm particles resulted in a smaller contact angle, smaller average agglomerate size, and lower galena recovery.

To achieve optimal particle size distribution and enhance fine-grained galena flotation recovery, thus promoting the effective recovery of micro-fine galena, it is essential to reduce the attrition of the grinding media on the mineral by adjusting the grinding parameters, such as by lowering the filling ratio. This minimizes galena overgrinding, lowers the −10 μm yield, and enhances flotation recovery. This provides some guidance for the industrial practice of reducing the yield of the −10 μm particle size fraction by adjusting the grinding process parameters.

Author Contributions: Conceptualization, G.G. and Y.W. (Yanhong Wang); methodology, G.G. and M.G.; formal analysis, Y.W. (Yanming Wu), S.Y. and Y.Y.; resources, G.G. and Y.W. (Yanhong Wang); data curation, M.G.; writing—original draft preparation, M.G.; writing—review and editing, G.G., Y.W. (Yanhong Wang) and J.C.; visualization, Q.L.; supervision, G.G.; project administration, G.G. and Y.W. (Yanhong Wang). All authors have read and agreed to the published version of the manuscript.

Funding: This research was funded by the National Natural Science Foundation of China, grant number 52074358.

Data Availability Statement: The data obtained and analyzed within the frame of this study will be made available by the corresponding author upon reasonable request.

Conflicts of Interest: Authors Yuankun Yang, Shengli Yu and Yanming Wu were employed by the company Zijin Mining Group Co., Ltd. The remaining authors declare that the research was conducted in the absence of any commercial or financial relationships that could be construed as a potential conflict of interest.

References

Nayak, A.; Jena, M.S.; Mandre, N.R. Beneficiation of Lead-Zinc Ores—A Review. *Miner. Process. Extr. Metall. Rev.* **2022**, *43*, 564–583. [CrossRef]

Zhang, X.W.; Yang, L.S.; Li, Y.H.; Li, H.R.; Wang, W.Y.; Ye, B.X. Impacts of lead/zinc mining and smelting on the environment and human health in China. *Environ. Monit. Assess.* **2012**, *184*, 2261–2273. [CrossRef] [PubMed]

Zhang, Z.; Xie, F.; Wang, W. A Review on Lead Extraction from Ore and Spent Lead Paste by Hydrometallurgical Processes. *Jom* **2024**, *76*, 5569–5588. [CrossRef]

Li, Q.K.; Du, S.Y.; Gu, G.H.; Wu, L.D.; Wang, Y.H. Utilizing the high intensity conditioning process with a multi-bladed impeller to enhance the flotation of fine-grained galena. *Part. Sci. Technol.* **2024**, *42*, 895–907. [CrossRef]

Zhang, X.L.; Wei, H.Q.; Han, Y.X.; Zhou, Z.Y.; Li, W.B. Effect of different mills on the fine grinding characteristics and leaching behaviour of gold ore. *Miner. Eng.* **2024**, *215*, 108800. [CrossRef]

Xiao, X.; Zhang, G.W.; Feng, Q.M.; Xiao, S.X.; Huang, L.L.; Zhao, X.; Li, Z.Q. The liberation effect of magnetite fine ground by vertical stirred mill and ball mill. *Miner. Eng.* **2012**, *34*, 63–69. [CrossRef]

Kumar, A.; Sahu, R.; Tripathy, S.K. Energy-Efficient Advanced Ultrafine Grinding of Particles Using Stirred Mills—A Review. *Energies* **2023**, *16*, 5277. [CrossRef]

8. Aleksandrova, T.N.; Orlova, A.V.; Taranov, V.A. Current Status of Copper-Ore Processing: A Review. *Russ. J. Non-Ferr. Met.* **20** *62*, 375–381. [CrossRef]
9. Du, Y.; Tong, X.; Xie, X.; Lu, Y.; Hua, Z.; Zhang, W. Effects of pyrite on the flotation behavior and electrochemical interaction different particle sizes of galena in the presence of diethyldithiocarbamate. *Sep. Sci. Technol.* **2021**, *56*, 2467–2474. [CrossRef]
10. Peng, Y.; Grano, S. Dissolution of fine and intermediate sized galena particles and their interactions with iron hydroxide colloi *J. Colloid Interface Sci.* **2010**, *347*, 127–131. [CrossRef]
11. Peng, Y.J.; Grano, S.; Fornasiero, D.; Ralston, J. Control of grinding conditions in the flotation of galena and its separation fro pyrite. *Int. J. Miner. Process.* **2003**, *70*, 67–82. [CrossRef]
12. Lu, Y.L.; Tong, X.; Xie, X.; Yang, B.; Hua, Z.B. Effect of particle size on the oxidation and flotation behavior of galena particl *Physicochem. Probl. Miner. Process.* **2019**, *55*, 208–216. [CrossRef]
13. Peng, Y.J.; Grano, S. Effect of iron contamination from grinding media on the flotation of sulphide minerals of different partic size. *Int. J. Miner. Process.* **2010**, *97*, 1–6. [CrossRef]
14. Guo, W.; Guo, K.Q.; Xing, Y.W.; Gui, X.H. A Comprehensive Review on Evolution Behavior of Particle Size Distribution Duri Fine Grinding Process for Optimized Separation Purposes. *Miner. Process. Extr. Metall. Rev.* **2024**. [CrossRef]
15. Miettinen, T.; Ralston, J.; Fornasiero, D. The limits of fine particle flotation. *Miner. Eng.* **2010**, *23*, 420–437. [CrossRef]
16. Tong, J.Q.; Wu, C.B.; Wang, Y.H.; Tian, J.K.; Li, Z.Y.; Xie, F.; Yao, X.; Zeng, G.S. Effect on Fine Particles Output Characteristics Ceramic Ball Grinding. *Minerals* **2023**, *13*, 1416. [CrossRef]
17. Wu, C.B.; Chen, Z.L.; Liao, N.N.; Zeng, C.; Wang, Y.H.; Tian, J.K. Enhancing the Grinding Efficiency of a Magnetite Second-Sta Mill through Ceramic Ball Optimization: From Laboratory to Industrial Applications. *Minerals* **2024**, *14*, 160. [CrossRef]
18. Matsanga, N.; Nheta, W.; Chimwani, N. A Review of the Grinding Media in Ball Mills for Mineral Processing. *Minerals* **2023**, *1* 1373. [CrossRef]
19. Shahbazi, B.; Jafari, M.; Parian, M.; Rosenkranz, J.; Chelgani, S.C. Study on the impacts of media shapes on the performance tumbling mills—A review. *Miner. Eng.* **2020**, *157*, 106490. [CrossRef]
20. Kutuk, S. Influence of milling parameters on particle size of ulexite material. *Powder Technol.* **2016**, *301*, 421–428. [CrossRef]
21. Chen, X.M.; Peng, Y.J. The effect of regrind mills on the separation of chalcopyrite from pyrite in cleaner flotation. *Miner. En* **2015**, *83*, 33–43. [CrossRef]
22. Hacifazlioglu, H.; Korkmaz, A.V. Performance comparison of stirred media mill and ball (BOND) mill in bauxite grinding. *Pa Sci. Technol.* **2020**, *38*, 404–408. [CrossRef]
23. Samanli, S.; Cuhadaroglu, D.; Kizgut, S. A Simulation Study of Laboratory Scale Ball and Vertical Stirred Mills. *Part. Part. Sys Charact.* **2010**, *26*, 256–264. [CrossRef]
24. Ma, S.J.; Li, H.J.; Shuai, Z.C.; Yang, J.L.; Deng, X.J.; Xu, W.Z. Research on Grinding Law and Grinding Parameters Optimization Polymetallic Complex Ores. *Minerals* **2022**, *12*, 1283. [CrossRef]
25. Yuan, C.F.; Wu, C.B.; Fang, X.; Liao, N.N.; Tong, J.Q.; Yu, C. Effect of Slurry Concentration on the Ceramic Ball Grindin Characteristics of Magnetite. *Minerals* **2022**, *12*, 1569. [CrossRef]
26. Jankovic, A. Variables affecting the fine grinding of minerals using stirred mills. *Miner. Eng.* **2003**, *16*, 337–345. [CrossRef]
27. Wang, C.; Deng, J.; Tao, L.M.; Sun, W.; Xiao, Q.F.; Gao, Z.Y. Enhanced flotation of chalcopyrite particles by grinding with sho cylinder media. *Miner. Eng.* **2022**, *188*, 107827. [CrossRef]
28. Partyka, T.; Yan, D. Fine grinding in a horizontal ball mill. *Miner. Eng.* **2007**, *20*, 320–326. [CrossRef]
29. Zhao, S.; Evans, T.M.; Zhou, X. Effects of curvature-related DEM contact model on the macro- and micro-mechanical behaviour of granular soils. *Geotechnique* **2018**, *68*, 1085–1098. [CrossRef]
30. Li, Q.; Zhou, J.; Wu, Y.; Yang, Y.; Wang, Y.; Gu, G.; Liu, B. Influence of the impeller structure and operating parameters of a hig intensity conditioning system on the flotation of fine-grained pyrite. *Miner. Eng.* **2025**, *228*, 109363. [CrossRef]
31. Yang, J.L.; Shuai, Z.C.; Zhou, W.T.; Ma, S.J. Grinding Optimization of Cassiterite-Polymetallic Sulfide Ore. *Minerals* **2019**, *9*, 13 [CrossRef]
32. Yang, J.L.; Deng, X.J.; Xu, W.Z.; Li, H.J.; Ma, S.J. Study on the Grinding Law of Ball Media for Cassiterite-Polymetallic Sulfide Or *Minerals* **2022**, *12*, 270. [CrossRef]
33. He, K.; Jia, M.P.; Sun, F.; Yang, Q.; Wu, B.; Li, X.B.; Meng, C. Simulation of Grinding Media Motion and Collisions in Wet-Operate Accelerator-Mill Using DEM Coupled with CFD. *Minerals* **2022**, *12*, 341. [CrossRef]
34. Puga, A.M.; Govender, N.; Rajamani, R.K. Verification of Polyhedral DEM with Laboratory Grinding Mill Experiments. *Kon Powder Part. J.* **2022**, *39*, 208–218. [CrossRef]
35. Cleary, P.W.; Owen, P. Effect of operating condition changes on the collisional environment in a SAG mill. *Miner. Eng.* **2019**, *13* 297–315. [CrossRef]
36. Fang, X.; Wu, C.B.; Liao, N.N.; Zhong, J.X.; Duan, X.Q.; Zhu, S.L.; Liu, A.J.; Xiao, K. Investigating the Influence of Medium Siz and Ratio on Grinding Characteristics. *Minerals* **2024**, *14*, 875. [CrossRef]

Xiaojing, Y.; Jinlin, Y.; Shaojian, M.; Runnan, G. Research on the characteristics of particles size for grinding products with a ball mill at low speed. *J. Phys. Conf. Ser.* **2021**, *2044*, 012040.

Xiaojing, Y.; Shaojian, M.; Jinlin, Y.; Pengyan, Z. Crushing characteristics of different minerals by cylindrical ball mill in cascading motion state. *J. Phys. Conf. Ser.* **2021**, *2044*, 012142.

Xiaojing, Y.; Shaojian, M.; Linling, J.; Runnan, G. Approach of establishing the grinding population balance kinetic model for cassiterite-polymetallic sulfide ore. *Ferroelectrics* **2021**, *578*, 66–80.

Li, Q.; Wu, L.; Yang, Y.; Gu, G.; Wang, Y. The relationship among pyrite fine grinding by stirred ball mill and tumbling ball mill, the properties of the product and the flotation behavior. *Miner. Eng.* **2025**, *229*, 109401. [CrossRef]

Gao, Z.Y.; Sun, W.; Hu, Y.H. Mineral cleavage nature and surface energy: Anisotropic surface broken bonds consideration. *Trans. Nonferrous Met. Soc. China* **2014**, *24*, 2930–2937. [CrossRef]

Gao, Z.Y.; Sun, W.; Hu, Y.H.; Liu, X.W. Surface energies and appearances of commonly exposed surfaces of scheelite crystal. *Trans. Nonferrous Met. Soc. China* **2013**, *23*, 2147–2152. [CrossRef]

Moon, K.S.; Fuerstenau, D.W. Surface crystal chemistry in selective flotation of spodumene ($LiAl[SiO_3]_2$) from other aluminosilicates. *Int. J. Miner. Process.* **2003**, *72*, 11–24. [CrossRef]

Alvarez, A.; Gutierrez, L.; Laskowski, J.S. Use of polyethylene oxide to improve flotation of fine molybdenite. *Miner. Eng.* **2018**, *127*, 232–237. [CrossRef]

Wan, H.; Qu, J.P.; Zhao, L.F.; Bu, X.Z. Improving the flotation performance of fine molybdenite using superhydrophobic magnetic carriers. *Miner. Eng.* **2025**, *224*, 109212. [CrossRef]

minerals

MDPI

Article

Research on the Dry Deep Flip-Flow Screening of Ilmenite and Its Pre-Throwing Tail Processing Technology

Wei Shi [1,2], Weinan Wang [1,3,4,*], Pengfei Mao [1], Xu Hou [1], Songxue Zhang [1] and Chenlong Duan [1,3,4,*]

1 School of Chemical Engineering and Technology, China University of Mining and Technology, Xuzhou 221116, China; david.shi@hewascreens.com (W.S.)
2 Jiangsu Hewa Screening Product Co., Ltd., Xuzhou 221116, China
3 International Joint Laboratory of Minerals Efficient Processing and Utilization, Ministry of Education, Xuzhou 221116, China
4 Jiangsu International University Cooperation Laboratory of Resource Low Carbon Processing & Utilization, China University of Mining and Technology, Xuzhou 221116, China
* Correspondence: wangwn@cumt.edu.cn (W.W.); clduan@cumt.edu.cn (C.D.); Tel.: +86-15852486321 (W.W.); +86-051683590178 (C.D.)

Abstract: Screening is a key step in the mineral process of ilmenite. As the grading particle size decreases, the phenomenon of clogged holes on the screening-plate intensifies, the screening environment deteriorates, and the screening effect deteriorates, seriously restricting subsequent sorting operations. This study proposes a 1 mm dry flip-flow screening method for ilmenite to achieve efficient deep classification of fine-grained materials. Firstly, a laser displacement testing system is used to study the dynamic characteristics of the flip-flow screen; based on the characteristics of different feed particle sizes, further research is conducted on the 1 mm dry flip-flow screening effect of ilmenite under different ratios of obstructive particles and difficult-to-screen particles. The 1 mm screening effect can reach 85.41%. Finally, the pre-throwing tailings process based on 1 mm multi-stage screening is put forward. This pre-throwing tail process has the characteristic of not using water and is suitable for sorting in arid, water deficient, and high-altitude frozen soil areas. It has important promotion and application value.

Keywords: ilmenite; flip-flow screen; dynamic characteristics

1. Introduction

Titanium, a crucial metal, serves as an essential constituent in the fabrication of high-strength and lightweight alloy materials [1,2]. Titanium predominantly occurs in minerals such as ilmenite. In recent years, with the continuous intensification of exploration and mining endeavors for ilmenite, and considering the extensive yet uneven distribution of ilmenite mineralization zones, ilmenite resources currently exhibit characteristics of low grade, complex associated minerals, and challenging development [3,4]. Thus, the development and utilization of ilmenite hold great significance.

Mineral processing can accomplish the enrichment and separation of useful minerals and is the fundamental technology for the efficient and clean utilization of mineral resources [5]. As the grade of ilmenite declines, particularly when the grain size of valuable components becomes finer, the raw ore needs to be thoroughly crushed and liberated through finer grinding. The finer the crushing and grinding particle size, the more readily useful minerals can be liberated, leading to higher efficiency in subsequent separation operations and a higher comprehensive recovery rate of useful minerals [6–8]. With the

https://doi.org/10.3390/min1503030

reduction in the raw ore particle size, it becomes necessary to further lower the lower limit of the classification particle size.

Screening is a pivotal step in the efficient and clean utilization of mineral resources. Its principle involves classifying materials into two or more different size fractions based on their particle size [9–11]. It mainly undertakes functions such as preparation screening, inspection screening, desliming screening, dewatering screening, desliming screening, and tailings dry discharge in mineral processing [12–18]. In the traditional dry screening process, the screened particles tend to become stuck on the screen surface and adhere to form aggregates, resulting in difficult stratification, the phenomenon of screen surface hole blinding intensifies, reduced processing capacity, and decreased screening efficiency [19–23]. Consequently, efficient dry deep screening technology has always been at the forefront and a research hotspot in the field of screening.

At present, a series of studies have been conducted both domestically and internationally on the screening of 6 mm and 3 mm. Yu et al. established a dynamic model of a vibrating flip-flow screen and verified through experiments that the vibrating flip-flow screen has good performance in screening sticky fine particles of 3 mm [24]. Chen et al. constructed a DEM-FEM coupled simulation model to study the motion of particles during the flip-flow screening process [25]. Akbari et al. optimized the separation performance technology by proposing a flip-flow screening technology based on airflow enhancement [26]. As the ore grade decreases, especially for minerals with finer disseminated grain size such as ilmenite, it is necessary to further break through the lower limit of dry screening to 1 mm. Through 1 mm screening, materials can be effectively separated, improving the efficiency of subsequent beneficiation processes, reducing the loss of useful minerals, and reducing the mechanical load on subsequent equipment. For the beneficiation process of ilmenite, fine ilmenite particles can be screened out in advance through 1 mm screening, avoiding them from entering the grinding stage and being over ground, which improves the recovery rate of ilmenite and effectively reduces the energy consumption of grinding equipment.

The flip-flow screen is based on the principle of double mass sub-resonance. Driven by the fixed frame and floating frame of the flip-flow screen, the polyurethane screen plate experiences alternating relaxation and tension, generating significant acceleration [27,28]. This can effectively facilitate the rapid loosening and stratification of ilmenite, thereby achieving efficient screening. At present, the flip-flow screen has basically achieved large-scale and efficient 3 mm dry screening of fine-grained materials, but there is relatively little research on the 1 mm high-efficiency dry screening technology and theory.

Consequently, this study puts forward a 1 mm dry flip-flow screening approach for ilmenite. Initially, the dynamic characteristics of the screening body during the screening process were investigated. Subsequently, further research was carried out on the 1 mm dry screening performance of ilmenite under different feed particle-size characteristics, validating the effectiveness of this method. Finally, the pre-throwing tailings process based on 1 mm multi-stage screening is put forward.

2. Experimental System and Materials

2.1. Flip-Flow Screening Experimental System

This study established the flip-flow screening experimental system for conducting 1 mm dry deep flip-flow screening experiments of fine-grained ilmenite. The system mainly includes key structures such as floating frame, fixed frame, excitation motor, elastic support, screen bracket, screen base, floating beam, fixed beam, flip-flow screen plate, shear spring, etc. In Figure 1A, the excitation motor is located above the fixed frame and two motors are arranged in parallel to generate vibration excitation for the screen; The elastic support is located above the bracket and connected to the fixed frame, mainly used for shock absorption. The fixed frame and floating frame are connected by a shear spring, as shown in Figure 1B. The schematic diagram of the structure of the flip-flow screen plate is shown in Figure 1C,D, with both ends connected to the floating beam and the fixed beam, respectively.

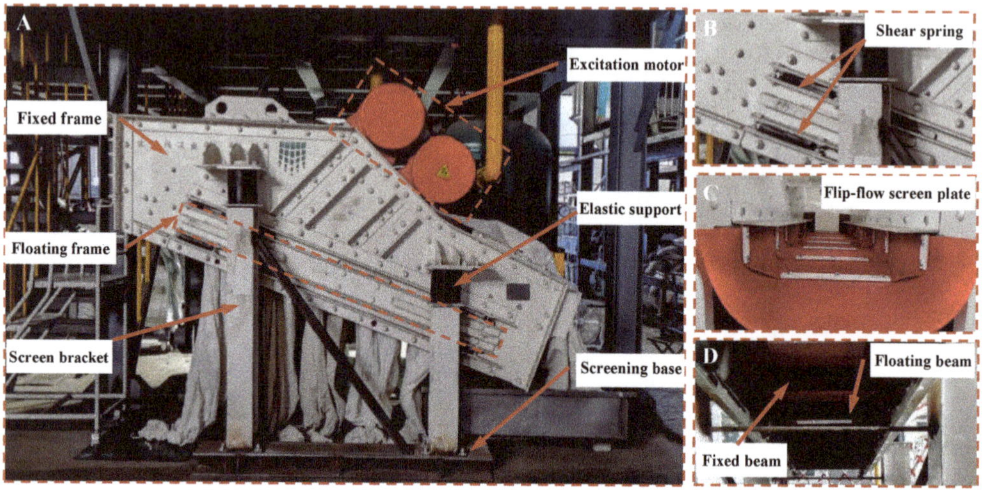

Figure 1. Flip-flow screening experimental system ((**A**). Front view of the flip-flow screen; (**B**). Partial diagram of shear spring; (**C**). Top view of screen-plate; (**D**). The bottom support beam of the flip-flow screen).

2.2. Evaluation Indicators for Flip-Flow Screening Effect

During the screening process of ilmenite, due to the inability of particles to pass through 100%, a portion of fine particles appeared in the oversized product; at the same time, due to the deformation of the elastic screen holes, some coarse particles appeared in the underflow products through the screen holes. Therefore, in the actual screening process, it is necessary to evaluate the screening performance of the flip-flow screen, and the specific calculation is as follows, and the definition of each parameter is shown in Table 1.

Table 1. The definition of each parameter [25].

Parameter	Definition	Unit	Parameter	Definition	Unit
U_c	Oversized fraction	%	O_f	Undersize fraction	%
$\gamma_{c,t}$	Theoretical yield of oversized product	%	$\gamma_{f,t}$	Theoretical yield of undersize product	%
γ_c	Actual yield of oversized product	%	γ_f	Actual yield of undersize product	%
M_c	Misplaced materials of coarse particles	%	M_f	Misplaced materials of fine particles	%
M_o	Total misplaced materials	%	η	Overall screening efficiency	%
E_c	Effective placement efficiency of coarse particles	%	E_f	Effective placement efficiency of fine particles	%

$$M_c = 100 \times \gamma_c \times O_f \tag{1}$$

$$M_f = 100 \times \gamma_f \times U_c \tag{2}$$

$$M_o = M_c + M_f \tag{3}$$

$$\eta = E_c + E_c - 100 \tag{4}$$

$$E_c = \frac{\gamma_c - M_c}{\gamma_{c,t}} \times 100 \tag{5}$$

$$E_f = \frac{\gamma_f - M_f}{\gamma_{f,t}} \times 100 \tag{6}$$

2.3. Characterization of ILMENITE

The screening test utilized dry ilmenite raw ore with particle size distribution divided into seven fractions: 6–4 mm, 4–3 mm, 3–2 mm, 2–1 mm, 1–0.71 mm, 0.71–0.5 mm, and 0.5–0 mm. Elemental analysis of the ilmenite ore was conducted using inductively coupled plasma optical emission spectroscopy (ICP-OES). The iron content was determined to be 13.046%, while titanium constituted 3.490% of the total composition.

3. Results and Discussion

3.1. Dynamic Characteristics and Spatial Motion Trajectory of Flip-Flow Screen

This study employed a laser displacement sensor based on the triangulation measurement method for dynamic testing. Specifically, the laser emitter projected the laser beam onto the surface of the tested screen-body. Subsequently, the scattered laser beam from the screen-body was captured by the camera within the receiver. The displacement of the object was calculated by measuring the distance between the laser beam and the screen-body. The flip-flow screen-body mainly consisted of a floating frame and a fixed frame, which were connected by a shear spring. During the screening process, the flip-flow motion of the screening body was achieved through the alternating movement of the floating frame and the fixed frame. Therefore, this study focused on the dynamic characteristics of the flip-flow screen floating frame and fixed frame.

Due to the fact that the floating frame mainly moves along the direction of material flow, the displacement in the other two directions remains almost unchanged. Therefore, the focus is on studying the time-varying characteristics of displacement in this direction. Firstly, experiments were carried out on the feeding area, middle area, and discharge area of the screen-body. The displacement changes of each measuring point on the floating frame of the screen-body over time were obtained, as presented in Figure 2. From the graph, it is evident that as time progresses, the displacement of each measuring point demonstrates periodic variations.

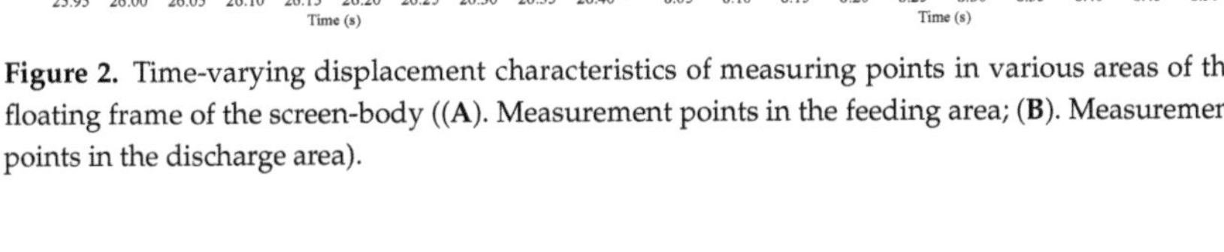

Figure 2. Time-varying displacement characteristics of measuring points in various areas of the floating frame of the screen-body ((**A**). Measurement points in the feeding area; (**B**). Measurement points in the discharge area).

In Figure 2A, the displacement range of the feeding end of the screen-body floating frame is from −5 mm to 7 mm, and the amplitude of the feeding end movement of the screen-body floating frame is 12 mm. In Figure 2B, the displacement range of the discharge end of the screen floating frame is from −7 mm to 6 mm, and the amplitude (the difference between the highest and lowest points of the movement) of the discharge end of the screen floating frame is 13 mm. The displacement magnitudes of the floating frame, in descending order, are: discharge area, feeding area.

Regarding the fixed frame of the screen-body, it primarily adheres to the direction of material flow and the normal direction perpendicular to the material flow. Herein, the x-axis direction denotes the movement direction along the material flow, while the y-axis direction represents the direction normal to the material flow. The time-varying displacement curves of the feeding and discharging ends of the screen-body's fixed frame along the x- and y-axes are depicted in Figure 3. As time elapses, the displacements of the feeding and discharging ends of the screen-body's fixed frame in the x and y directions exhibit periodic fluctuations.

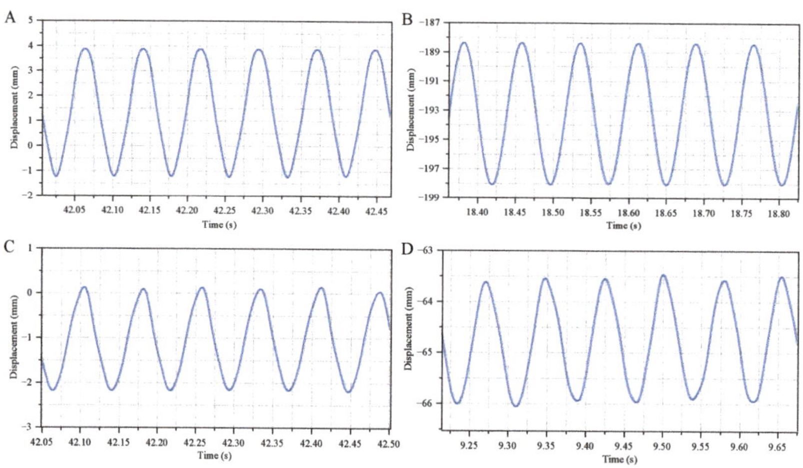

Figure 3. Time-varying displacement curves and motion trajectories of measuring points in various areas of the fixed frame of the screen-body ((**A**). Measurement points in the feeding area along the x-axis direction; (**B**). Measurement points in the feeding area along the y-axis direction; (**C**). Measurement points in the discharge area along the x-axis direction; (**D**). Measurement points in the discharge area along the y-axis direction).

In the feeding area illustrated in Figure 3A,B, the displacement range of the screen-body's fixed frame along the x-axis is from −1.25 mm to 4 mm, and the displacement range along the y-axis is from −198 mm to 188 mm. Evidently, the variation in displacement amplitude along the y-axis is significantly greater than that along the x-axis. In Figure 3C,D, within the discharge area, the displacement range of the screen-body's floating frame along the x-axis is from −2.25 mm to 0.25 mm, and the displacement range along the y-axis is from −66 mm to 62 mm. Along the material flow direction, the displacement amplitude of the feeding area of the screen-body's fixed frame is larger than that of the discharging area. Along the normal direction of the material flow, the displacement amplitude of the discharge area is greater than that of the feeding area. As can be observed from the figure below.

3.2. Influence of Particle-Size Distribution on the 1 mm Flip-Flow Screening Performance of Ilmenite

The foregoing research successfully derived the dynamic characteristics of the flip-flow screen. These characteristics enabled the stable operation of the screen-body and further facilitated the 1 mm screening experiment research on ilmenite. During the flip-flow screening process, the particle-size distribution characteristics of different materials played a crucial role. They affected both the probability of particles coming into contact with the screen-mesh and the probability of particle groups passing through the screen, ultimately influencing the final screening efficiency. In particular, for materials with particle sizes close to the screen aperture, the impact of particle-size distribution characteristics on the flip-flow screening performance was even more pronounced. Based on these findings, this section focused on exploring how the particle-size distribution characteristics of materials with sizes close to the screen aperture affected the screening performance of ilmenite, aiming to provide guidance for practical production.

Regarding particles with sizes close to the screen aperture, in combination with traditional particle-size classification methods in the screening process, and according to the screening difficulty of particles, they were classified into four categories: coarse particles (6–4 mm, 4–3 mm, 3–2 mm, 2–1.5 mm), hindering-particles (1.5–1 mm), difficult-to-screen particles (1–0.71 mm), and easy-to-screen particles (0.71–0.5 mm, 0.5–0 mm). Among these, hindering-particles and difficult-to-screen particles were the particle-size grades close to the screen aperture size. Meanwhile, the external moisture content of the materials was tested before and after each experiment. Due to the evaporation of moisture during the experiment, there was a certain difference in the external moisture content before and after the experiment, and the external moisture content of the materials was tested to be between 1.1% and 1.6%.

3.2.1. Influence of Particle-Size Distribution of Hindering Particle on the 1 mm Flip-Flow Screening Performance of Ilmenite

Given that the material within the 2–1 mm particle-size range was predominantly concentrated in the 1.5–1 mm range, and the quantity of material within the 2–1.5 mm particle-size range was extremely limited. For the sake of statistical convenience and calculation accuracy, the particle-size range of 2–1 mm was adopted to represent obstructive particles. The core objective of this study was to examine the influence of the gradation characteristics of hindering-particles and difficult-to-screen particles on the deep screening efficiency of the dry screening process.

The variations in the yield and weighted average particle size of the products obtained from dry deep flip-flow screening under different particle-size distribution conditions were presented in Figure 4. In Figure 4A, as the proportion of hindering-particles rose, the yield of the screened product showed a gradual increase, spanning from 58.47% to 67.32%. Concurrently, the weighted average particle size gradually diminished, ranging from 2.5 mm to 2.91 mm. In Figure 4B, with the gradual increase in the proportion of hindering-particles, the yield of the screened product gradually declined, within the range of 32.68% to 41.53%. Meanwhile, the weighted average particle size gradually increased, varying from 0.71 mm to 0.75 mm.

In Figure 5, as the proportion of hindering-particles gradually increased from 10% to 30%, the upper limit rate of the dry deep flip-flow screening product gradually increased, ranging from 1.68% to 5.87%, and the lower limit rate gradually decreased, ranging from 10.34% to 13.03%.

The distribution rate of materials with different particle sizes in the screened product under different hindering particle-size distribution conditions was depicted in Figure 6. As the proportion of hindering-particles increased, the distribution curve gradually shifted to the left. Regarding the distribution rate of materials of the same particle size, the larger the proportion of hindering-particles in the screened feed, the greater the amount of material of that particle size in the screened product. In Figure 6B, as the proportion of hindering-particle rose, their particle-size distribution gradually decreased, ranging from 1.23 mm to 1.25 mm. Simultaneously, the separation accuracy gradually declined, with a range from 0.19 mm to 0.22 mm.

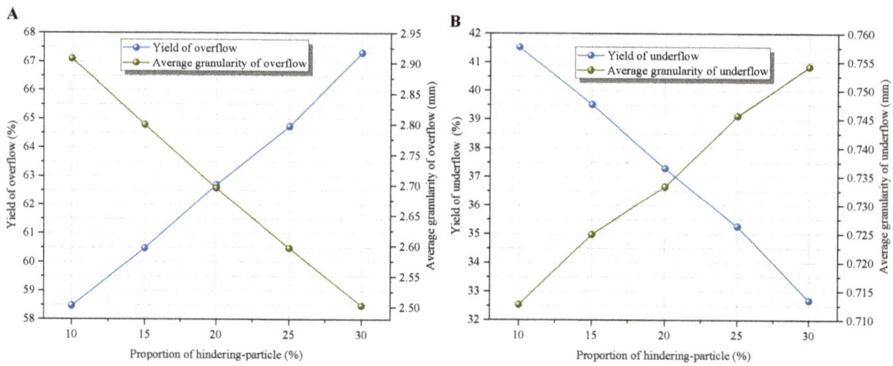

Figure 4. Yield and weighted average particle size of the dry deep flip-flow screening products under different proportions of hindering-particle. ((**A**). Overflow; (**B**). Underflow).

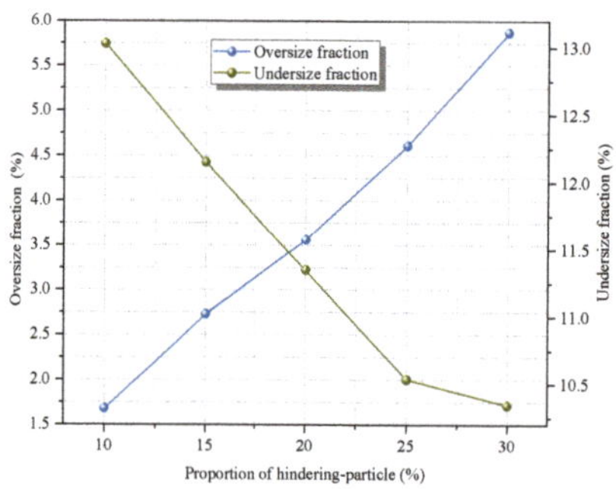

Figure 5. Oversize fraction and undersize fraction of the dry deep flip-flow screening products under different proportions of hindering-particle.

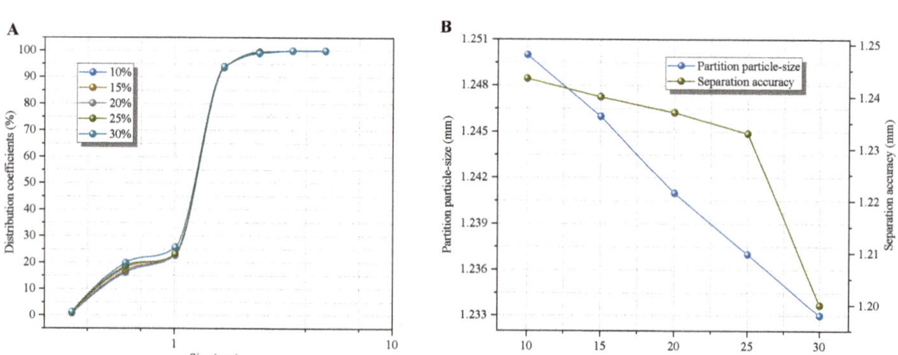

Figure 6. Distribution curve and separation accuracy of the dry deep flip-flow screening products under different proportions of hindering-particle. ((**A**). Distribution curve; (**B**). Separation accuracy).

The alterations in the screening efficiency and total mismatch content of the flip-flow screen under different particle size distribution conditions were presented in Figure 7. In Figure 7A, as the proportion of obstructive particles increased, both the positive blending ratios of materials with particle sizes less than 1 mm in the undersized product and materials with particle sizes greater than 1 mm in the oversized product decreased. Meanwhile, the overall dry deep flip-flow screening efficiency also declined from 82.92% to 78.46%. In Figure 7B, as the proportion of hindering particles gradually rose, the total mismatch content of dry deep flip-flow screening gradually increased from 8.32% to 8.88%.

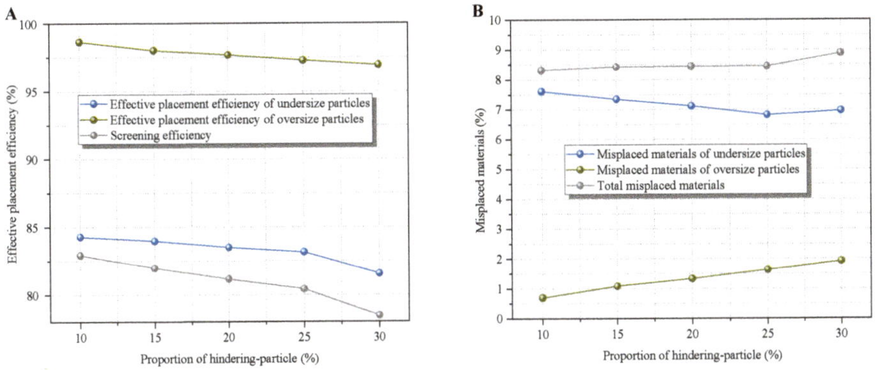

Figure 7. Screening efficiency and total misplaced materials of the dry deep flip-flow screening under different proportions of hindering-particle. ((**A**). Screening efficiency; (**B**). Misplaced materials).

3.2.2. Influence of Particle-Size Distribution of Particle Difficult to Be Screened on the 1 mm Flip-Flow Screening Performance of Ilmenite

The variations in the yield and weighted average particle size of the products from dry deep flip-flow screening under different difficult particle-size grading conditions were presented in Figure 8. In Figure 8A, as the proportion of difficult-to-screen particles increased, the yield of the screened product initially decreased and then increased. Meanwhile, the weighted average particle size of the screened material gradually diminished. In Figure 8B, the yield of the screened material first increased and then decreased. Correspondingly, its weighted average particle size first rose and then fell.

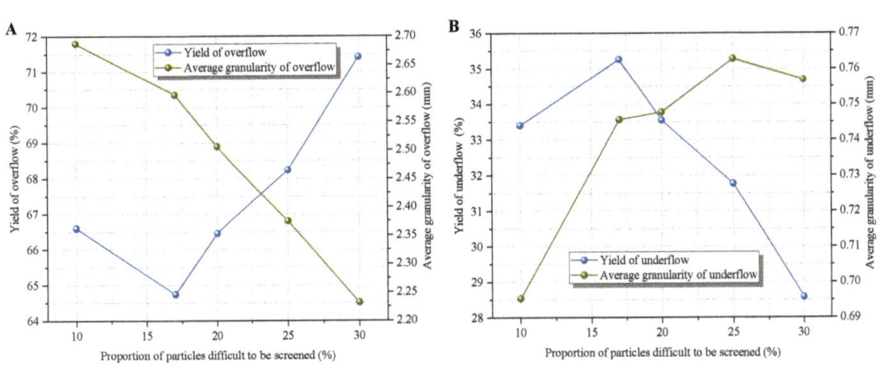

Figure 8. Yield and weighted average particle size of the dry deep flip-flow screening products under different proportions of difficult-to-screen particles. ((**A**). Overflow; (**B**). Underflow).

In Figure 9, as the proportion of difficult-to-screen particles increased, the upper limit rate of dry deep screening products gradually decreased within the range of 10% to

30%, specifically from 4.06% to 5.88%. Meanwhile, as the proportion of difficult-to-screen particles gradually rose, the lower limit rate of dry deep flip-flow screening products gradually increased, ranging from 6.17% to 31.65%.

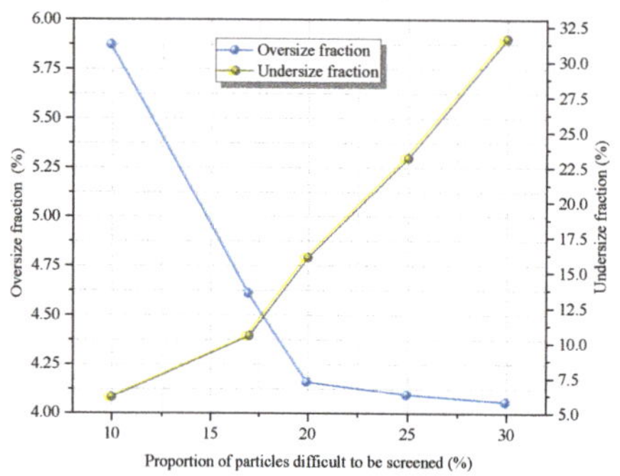

Figure 9. Oversize fraction and undersize fraction of the dry deep flip-flow screening products under different proportions of difficult-to-screen particles.

Evidently, the trend of the lower limit rate was more notable. This was mainly because as the proportion of difficult-to-screen particles increased, their particle size was slightly smaller than the screen hole size, classifying them as fine-grained materials. Moreover, the lower limit rate reflected the scenario where fine-grained materials failed to pass through the screen. Consequently, as the proportion of such materials increased, more fine-grained materials faced difficulties in passing through the screen. On the other hand, this also demonstrated that the lower limit rate was more sensitive to the changes in difficult-to-screen particles.

The distribution rate of each particle size in the flip-flow screening product under different particle-size distribution conditions was presented in Figure 10. As the proportion of difficult-to-screen particles increased, the distribution curve of the flip-flow screening products gradually shifted to the left. Regarding the distribution rate of materials of the same particle size, the larger the proportion of difficult-to-screen particles in the screened feed, the greater the amount of material of that particle size in the screened product. In Figure 10B, as the proportion of difficult-to-screen particles rose, the particle-size distribution gradually decreased, ranging from 0.89 mm to 1.28 mm. Simultaneously, the separation accuracy gradually declined, with a range from 0.16 mm to 0.38 mm.

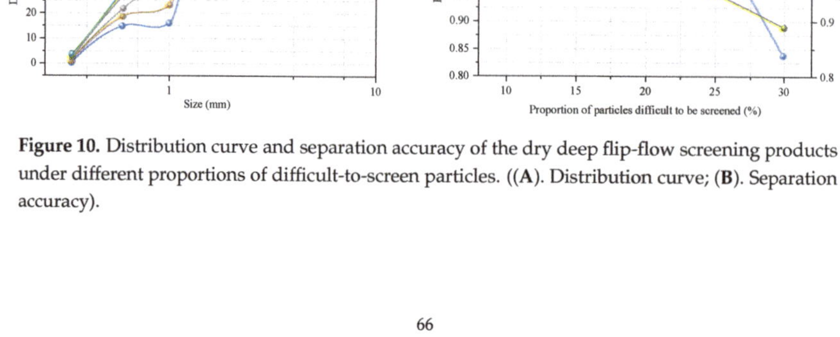

Figure 10. Distribution curve and separation accuracy of the dry deep flip-flow screening products under different proportions of difficult-to-screen particles. ((**A**). Distribution curve; (**B**). Separation accuracy).

The alterations in the screening efficiency and total mismatch content of the dry deep flip-flow screen under different particle-size distribution conditions were depicted in Figure 11. In Figure 11A, as the proportion of difficult-to-screen particles increased, the positive blending ratio of materials smaller than 1 mm in the undersized product decreased, while that of materials larger than 1 mm in the oversized product increased. Consequently, the overall screening efficiency of the screen gradually declined from 85.41% to 52.49%. In Figure 11B, as the proportion of difficult-to-screen particles rose, the total mismatch content of the dry deep flip-flow screen gradually increased from 6.07% to 23.76%.

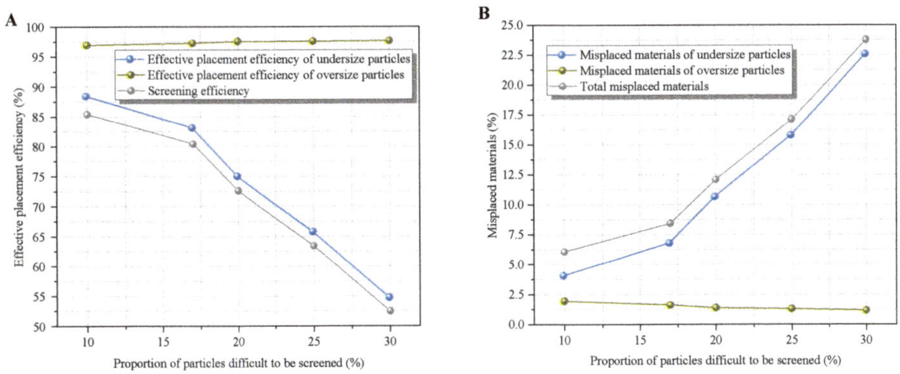

Figure 11. Screening efficiency and total misplaced materials of the dry deep flip-flow screen. ((**A**). Screening efficiency; (**B**). Misplaced materials).

3.3. *Pre-Throwing Tailings Process for Ilmenite Based on 1 mm Multi-Stage Dry Screening*

The aforementioned study validated the feasibility of the 1 mm dry screening technology for ilmenite from two perspectives: the dynamic characteristics of flip-flow screening and the impact of the feed particle-size characteristics on the 1 mm screening efficiency. Consequently, a tail-discarding sorting process for titanium-iron ore based on 1 mm dry multi-stage screening was put forward, as depicted in Figure 12.

The ilmenite raw ore was initially crushed, and the crushed product then proceeds to the first screening operation. The first screening utilized double-layer screen. The upper screen aperture was set at 12 mm, which could be adjusted in accordance with the crushing process requirements. The lower screen aperture was 1 mm. The +12 mm coarse product was returned to the crushing operation, while the 12–1 mm product was fed into the dry grinding operation, which employed a high-pressure roller mill. The grinding products were then directed to the second screening operation. The second screening used a 1 mm single-layer screen. The oversized product on the +1 mm was re-introduced into the dry grinding operation. The undersize product screened by the −1 mm was combined with the undersize product from the first screening and then enters first and second dry magnetic separation operations. Through two-stage strong magnetic separation and weak magnetic separation operations, titanium rough concentrate, iron rough concentrate, and tailings could be obtained. Moreover, the dry pre-tailing disposal of ilmenite tailings was successfully achieved.

This process reduced the number of fine particles entering the high-pressure roller mill through 1 mm multi-stage dry screening, which could effectively avoid excessive crushing of materials; at the same time, it reduced grinding energy consumption and achieved energy conservation and emission reduction. This pre-throwing tail process had the characteristic of not using water and was suitable for sorting in arid, water deficient, and high-altitude frozen soil areas. It has important promotion and application value.

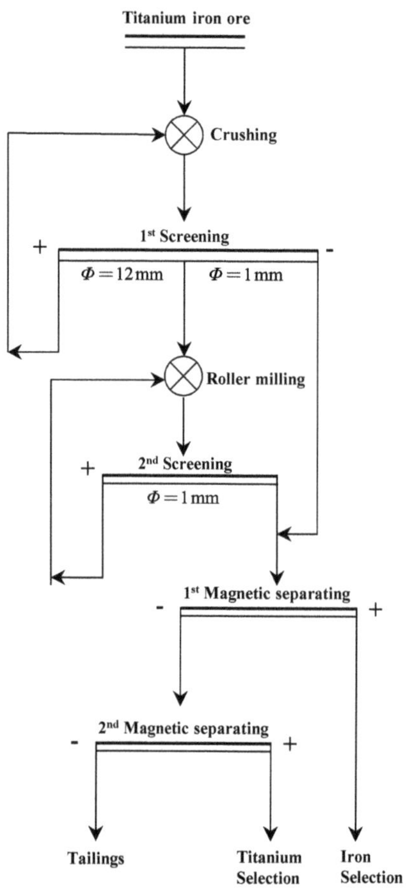

Figure 12. Pre-throwing tailings process for ilmenite based on 1 mm multi-stage screening.

4. Conclusions

As the lower limit of particle-size decreased, the difficulty of dry deep screening also increased. In response to the industry challenge of reducing the lower limit of screening from 6 mm and 3 mm to 1 mm, this study proposes a 1 mm dry flip-flow screening method for ilmenite to achieve effective deep classification of fine-grained materials and verifies the feasibility of this method by examining both the dynamic characteristics of the screening machine and its performance under different particle-size distributions. Based on this, a pre-throwing tailings process based on 1 mm multi-stage screening is proposed.

1. The floating frame has undergoes cyclic reciprocating motion along the direction of the material flow, with displacement amplitudes ranging from 12 to 13 mm. The screen-body experiences periodic, approximately linear motion.
2. As the proportion of hindering particles increased, the 1 mm screening efficiency also decreased from 82.92% to 78.46%. As the proportion of difficult-to-screen particles increased, the 1 mm screening efficiency also decreased from 85.41% to 52.49%.
3. Pre-throwing tailings separation process based on 1 mm multi-stage screening is proposed. This process effectively avoids excessive crushing of materials and reduces grinding energy consumption.

Author Contributions: Conceptualization, W.W. and W.S.; methodology, P.M.; software, S.Z.; validation, W.S. and C.D.; formal analysis, W.S.; investigation, X.H.; resources, W.W.; data curation, P.M.; writing—original draft preparation, W.S.; writing—review and editing, W.W.; supervision, C.D.; project administration, W.W. and C.D; funding acquisition, W.W. and C.D. All authors have read and agreed to the published version of the manuscript.

Funding: The research work is financially supported by National Natural Science Foundation of China (52404303), the Funded by Basic Research Program of Jiangsu (BZ2024046), the Serving "343" Industry Development Project in Universities in Xuzhou (gx2023004), and the China Postdoctoral Science Foundation (2024M753530).

Data Availability Statement: Data will be made available on request.

Conflicts of Interest: Wei Shi is an employee of Jiangsu Hewa Screening Product Co., Ltd. The paper reflects the views of the scientists and not the company.

eferences

El Khalloufi, M.; Drevelle, M.O.; Soucy, G. Titanium: An Overview of Resources and Production Methods. *Minerals* **2021**, *11*, 1425. [CrossRef]

Sengottuvelan, A.; Balasubramanian, P.; Will, J.; Boccaccini, A.R. Bioactivation of titanium dioxide scaffolds by ALP-functionalization. *Bioact. Mater.* **2017**, *2*, 108–115. [CrossRef] [PubMed]

Zhang, S.; Rao, M.; Xiao, R.; You, J.; Li, G.; Jiang, T. Enrichment of Nb and Ti from carbonatite pyrochlore ore via calcining-slaking followed by gravity separation. *Int. J. Min. Sci. Technol.* **2022**, *32*, 615–626. [CrossRef]

Xie, S.; Hu, Z.; Lu, D.; Zhao, Y. Dry Permanent Magnetic Separator: Present Status and Future Prospects. *Minerals* **2022**, *12*, 1251. [CrossRef]

Zheng, X.; Li, X.; Du, L.; Zhou, S.; Wang, Y.; Lu, D. Clean and short flowsheet of recovering ultrafine ilmenite by high gradient magnetic separation coupling with magnetic fluid. *Miner. Eng.* **2025**, *222*, 109156. [CrossRef]

Dong, S.; Li, T.; Yu, J.; Wei, Q.; Ren, X. A strategy for treatment of low-grade ore: Efficient separation and purification of iron. *Process Saf. Environ. Prot.* **2024**, *191*, 1313–1323. [CrossRef]

Zhu, K.; Wei, Q.; Li, H.; Ren, X. Solvent extraction of titanium from ilmenite hydrochloric acid leachate: Optimization and investigation of extraction reactions of all contained metal ions. *Miner. Eng.* **2022**, *186*, 107744. [CrossRef]

Yuan, Z.; Zhao, X.; Lu, J.; Lv, H.; Li, L. Innovative pre-concentration technology for recovering ultrafine ilmenite using superconducting high gradient magnetic separator. *Int. J. Min. Sci. Technol.* **2021**, *31*, 1043–1052. [CrossRef]

Han, J.; Zhang, J.; Feng, W.; Chen, X.; Zhang, L.; Tu, G. A Clean Process to Prepare High-Quality Acid-Soluble Titanium Slag from Titanium Middling Ore. *Minerals* **2019**, *9*, 460. [CrossRef]

10. Wang, Y.; Wen, S.; Feng, Q.; Ye, Z.; Wang, M. Mineral phase reconstruction behavior of direct reduction and smelting titanium slag at high temperature and slow cooling. *Rare Met.* **2015**, *34*, 440–444. [CrossRef]

11. Ashika, S.A.; Balamurugan, S.; Fathima, T.K.S. Phase evolution of nanocrystalline Mn-based oxides screened under different calcination temperatures using different precursors for proficient application in near infrared pigmentation. *Adv. Powder Technol.* **2024**, *35*, 104482. [CrossRef]

12. Guerreiro, F.S.; Gedraite, R.; Ataíde, C.H. Residual moisture content and separation efficiency optimization in pilot-scale vibrating screen. *Powder Technol.* **2016**, *287*, 301–307. [CrossRef]

13. Li, Y.; Peng, Z.; Wang, Z.; Zhu, Y.; Xie, K. Study on the Properties of Vanadium Pellets Extracted from Vanadium Titanium Magnetite Concentrate by Calcium Roasting and Acid Leaching. *Minerals* **2023**, *13*, 399. [CrossRef]

14. Costa, B.S.; Bergerman, M.G.; Júnior, H.D. Comparing the Performance of Hydrocyclones and High-Frequency Screens in an Industrial Grinding Circuit: Part I—Size Separation Assessments. *Minerals* **2024**, *14*, 707. [CrossRef]

15. Lin, D.; Wang, X.; Xu, N.; Zuo, W.; Liang, Z. A Method for Stabilizing the Vibration Amplitude of a Flip-Flow Vibrating Screen Using Piecewise Linear Springs. *Minerals* **2024**, *14*, 406. [CrossRef]

16. Xu, N.; Wang, X.; Lin, D.; Zuo, W. Numerical Simulation and Optimization of Screening Process for Vibrating Flip-Flow Screen Based on Discrete Element Method–Finite Element Method–Multi-Body Dynamics Coupling Method. *Minerals* **2024**, *14*, 278. [CrossRef]

17. Hu, Z.; Lu, D.; Wang, Y.; Zheng, X.; Zhang, Y. A novel pneumatic dry high-intensity magnetic separator for the beneficiation of fine-grained hematite. *Powder Technol.* **2024**, *433*, 119216. [CrossRef]

18. Xiong, X.; Niu, L.; Gu, C.; Wang, Y. Vibration characteristics of an inclined flip-flow screen panel in banana flip-flow screens. *J. Sound Vib.* **2017**, *411*, 108–128. [CrossRef]

19. Sun, T.; Wu, B.; Zhang, H.; Han, Y.; Liu, H.; Zhang, Y. Discrete element simulation of particle flow and separation in a vibratir screen with variable rectangular hole screen. *Powder Technol.* **2024**, *434*, 119305. [CrossRef]
20. Fu, J.; Zhang, J.; Liu, F. Enhanced sieving mechanism of novel cleaning screen and investigation of particle movement character tics on the screen. *Powder Technol.* **2024**, *431*, 119043. [CrossRef]
21. Makinde, O.A.; Ramatsetse, B.L.; Mpofu, K. Review of vibrating screen development trends: Linking the past and the future mining machinery industries. *Int. J. Miner. Process.* **2015**, *145*, 17–22. [CrossRef]
22. Asghari, M.; Noaparast, M.; Chegeni, M.J. Improving the roller screen efficiency to classify green iron pellets using DE simulation, novel roll design and implementing banana configuration. *Adv. Powder Technol.* **2024**, *35*, 104675. [CrossRef]
23. Gong, S.; Wang, C.; Guo, J.; Qiao, Z.; Zhao, G.; Fan, J.; Xu, N.; Wang, X. Dynamic analysis and optimization of the couplin system of vibrating flip-flow screen and material group. *Symmetry* **2024**, *16*, 913. [CrossRef]
24. Yu, C.; Pu, K.; Geng, R.; Qiao, D.; Lin, D.; Xu, N.; Wang, X.; Li, J.; Gong, S.; Zhou, Q. Comparison of flip-flow screen and circul vibrating screen vibratory sieving processes for sticky fine particles. *Miner. Eng.* **2022**, *187*, 107791. [CrossRef]
25. Chen, B.; Yu, C.; Gong, S.; Wang, X. Dynamic characteristics of LIWELL flip-flow screen panel and particle movement. *Chem. En Sci.* **2021**, *245*, 116853. [CrossRef]
26. Akbari, H.; Ackah, L.; Mohanty, M. Performance optimization of a new air table and flip-flow screen for fine particle d separation. *Int. J. Coal Prep. Util.* **2020**, *40*, 581–603. [CrossRef]
27. Cao, P.; Tang, J.; Xiong, X.; Niu, L. Synchronous control strategy of double excitation motors inertial flip-flow screen. *ISA Trar* **2022**, *131*, 566–578. [CrossRef] [PubMed]
28. Li, L.; Xing, Z.; Hao, H.; Yu, Z. Dynamic analysis and experimental study of the flip-flow vibrating screen(FFVS)'s sieve pan based on membrane vibration theory. *Powder Technol.* **2024**, *439*, 119745. [CrossRef]

minerals

MDPI

ticle

ffect of Wall Roughness in the Middle Zone of Spiral :oncentrator on the Flow Field Evolution of Hematite–Quartz lurry and Particle Separation Behaviour

ıuling Gao *, Xiaohong Zhou *, Bochao Li, Qian Wang and Chunyu Liu

School of Resources and Civil Engineering, Northeastern University, Shenyang 110819, China; 2490134@stu.neu.edu.cn (B.L.); wangqian@stu.neu.edu.cn (Q.W.); 2390136@stu.neu.edu.cn (C.L.)

* Correspondence: gaoshuling@mail.neu.edu.cn (S.G.); 2310468@stu.neu.edu.cn (X.Z.); Tel.: +86-137-0981-8446 (S.G.); +86-135-1608-6011 (X.Z.)

Abstract: The spiral concentrator is usually the first operation in the combined process of iron ore beneficiation. The industrial separation index decreases as the trough surface undergoes increased wear. A combination of surface roughness measurement and numerical experimental methods is utilized to systematically investigate the effect of wall roughness in the middle zone on the evolution of the flow field of a slurry consisting of hematite, quartz and water in a spiral concentrator. The radial migration and distribution characteristics of hematite and quartz particles are analysed, and the separation indexes are further evaluated. The results show that an increase in wall roughness (K_s value) in the middle zone has been shown to decrease the depth of slurry flow, the velocity and radial flux of secondary flow in the inner and middle zones of the trough, and to narrow the space of inward flow. The variation in hydrodynamic parameters is particularly pronounced as the K_s value increases from 0.1 mm to 0.2 mm, resulting in a significant reduction in the space available to the separation fluid and an observable interruption in the inward flow. The migration tendency of hematite and quartz particles to the inner trough is reduced depending on the flow field parameters, and their enrichment zones are both shifted outward. The migration amount and distance of particles show apparent differences in density and size. The separation indexes decrease slightly as the wall roughness (K_s value) in the middle zone increases in the 0.01 to 0.1 mm range, but the iron grade of concentrate decreases significantly, and the separation effect worsens as the K_s value increases from 0.1 mm to 0.2 mm. The separation effect of hematite and quartz particles in the spiral concentrator is influenced by the comprehensive interaction of feed size and wall roughness in the middle zone. The results of this study provide a theoretical basis for the selection of trough material, surface structure design and the production process control of the spiral concentrator.

Keywords: spiral concentrator; wall roughness; slurry flow field; radial flux; particle separation

1. Introduction

The spiral concentrator is a kind of thin-film flow gravity separation equipment which is widely used in ore separation and coal washing due to its simple manufacture, low cost and environmental protection. In the rotating flow field of the spiral concentrator, solid particles make complex three-dimensional movements based on the combined effect of centrifugal inertia force and gravity. Ultimately, separation is achieved based on differences in motion behaviour. The intrinsic reason for the separation is based on the differences in the physical properties of the particles, while the extrinsic reason is based on the characteristics of the flow field in the separation space. Over the past decades, researchers have conducted many physical and numerical experiments to investigate the effects of structural and operational factors on this complex flow pattern and particle separation [1–8]. However, most of these studies are based on an assumption of smooth walls, and do not consider actual wall conditions.

The liquid flow on the surface of the spiral concentrator involves laminar, turbulent and transitional flow, which is characterised by the coexistence of multiple flow regimes. Wall roughness is one of the most important properties of the wall surface, and the separation of particles in the spiral concentrator arises from the interaction between the slurry and the wall surface, resulting in the main flow and secondary flow [9]. Experimental studies of turbulent boundary layers have shown that wall roughness significantly affects the turbulence structure, boundary layer thickness and even the velocity distribution of the main flow streams [10,11]. Variations in fluid flow characteristics and turbulence further affect the velocity and distribution of particles, which can lead to differences in the collision, attachment and separation processes of different particles [12–15]. Therefore, the wall roughness has an important effect on the liquid flow and particle movement process in the spiral concentrator. In iron ore production, the spiral concentrator is usually used as the first operation of the combined separation process. Due to the constraints of the previous grading operation and the impact of equipment conditions, the bottom of the trough body is often attached to the material layer or even piles up and breaks the flow, seriously affecting the separation performance of the equipment. Nonetheless, owing to the discrepancy in processing and production technology, coupled with the abrasion of ore particles in the production process, the wall roughness of the spiral concentrator is not quantitative and complex. Inadequate control readily results in a reduction in production indicators [16].

In fact, the roughness usually occurs in a random pattern because it contains different sizes and irregularities that cause the effect of surface roughness to increase gradually, and each rough element becomes active when it reaches a critical Reynolds number. Therefore it is quite difficult to make detailed measurements or numerical simulations of the flow, especially near the wall roughness [17,18]. Roughness affects many turbulent flows in engineering and the natural sciences, and understanding its effects is important for predicting flow behaviour in most practical applications. The two main numerical simulation methods for assessing the effect of surface roughness on drag are the equivalent sand roughness model (K_s) and the discrete element model (DEM). The equivalent sand roughness model is empirical. Rough surfaces with different characteristics are compared with data from the flow in a pipe to assign an equivalent grit roughness height value to the rough surface. It overcomes the need for fine mesh resolution of each roughness element and has been the most widely used roughness model due to its simplicity and versatility [19,20].

In recent years, research work on wall roughness for spiral concentrators and related cyclones has opened up more possibilities for improving their separation performance [9,21–25]. The liquid flow pattern in the spiral concentrator shows thin inner and thick outer film-type flow along the surface of the trough, and the hydrodynamic parameters at the bottom of the trough are more sensitive to the change of wall roughness. Arnold et al. [21] found that, when the height of the protrusion in the middle of the trough surface reaches 5%–10% of the fluid depth, that is sufficient to cause fluid splitting, significantly affecting the continuity of the secondary circulation, whereas the free surface is not sensitive to the trough protrusion. Romeijn et al. [22] further found that the change in wall roughness also had no significant effect on either the free surface shape of the water in the tank surface or the bubble line. These studies have improved the intuitive understanding of the effect of wall roughness on the flow field in spiral concentrators, and also laid a hydrodynamic foundation for the in-depth revelation of the mechanism of particle–fluid interaction. In addition, the above effects of wall roughness on the flow field are mainly focused on the outer edge of the spiral concentrator and the thicker flow layer in the spiral channel [21,22]. However, it is a fact that the main zones, where stacking and sanding occur, are the inner and middle zones of the thin-flow film distribution in the spiral concentrator. In addition, the middle zone of the trough surface is often the easy mixing zone of low-density coarse particles and high-density particles in the spiral concentrator separation process, and the geometric parameter of the middle section has a significant effect on the particle separation process [26–28]. On this basis, it is necessary to systematically study the effect of the wall roughness of the middle trough surface on the slurry flow characteristics and particle separation, as well as the intrinsic correlation between the two. This can provide a basis for improving the separation efficiency of the spiral concentrator.

Based on the above research and application background, a spiral flow field and particle phase calculation model with higher accuracy, established in the previous stage, is used [26,29]. The hematite–quartz slurry system is taken as the research object, and experiments are carried out through wall roughness setting and rationality verification. The effect of wall roughness on the longitudinal evolution of fluid parameters, such as the spreading of the slurry flow layer, the spatial distribution of the secondary flow, the fluid velocity and the radial flux, is systematically investigated in the middle trough of the spiral concentrator. The radial migration behaviour and distribution characteristics of hematite and quartz are further examined. The hematite–quartz segregation index was evaluated, and changes in particle motion with fluid evolution are analysed. Suitable wall roughness conditions are explored. Mechanisms of the hydrodynamic influence of wall roughness in the middle zone on the process and outcome of particle separation are identified.

2. Methodology

2.1. Geometric Modelling and Meshing

A laboratory cubic parabolic spiral concentrator is taken as the object of study. Its sectional curve is shown in Figure 1, and other geometric parameters are shown in Table 1. The same structural parameters are taken for numerical simulation and validation experiments.

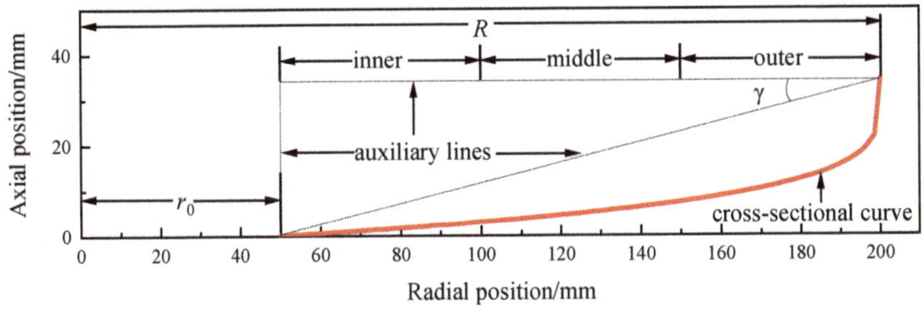

Figure 1. Cross-sectional profile of the spiral concentrator.

The entire spiral concentrator computational domain is shown in Figure 2a, which can be divided into inlet surface, outlet surface, upper surface and wall surface, and the wall surface includes the inner and outer side walls and the bottom wall of the trough. Figure 2 further shows an enlarged view of the 1.25-turn computational domain mesh, an enlarged top view of the computational domain mesh, and the mesh at the cross-section.

Based on the mesh-independence test, the fluid computational domain is divided into 8.71×10^5 hexahedral meshes. The bottom wall mesh is also locally encrypted to capture the complex flow at the bottom wall with a minimum bottom wall mesh height of 0.1 mm and a minimum y^+ value of about 2.

Table 1. Structural parameters of the spiral concentrator.

Structural Parameters	Value
Outer radius R/mm	200
Inner radius r_0/mm	50
Pitch of spiral/mm	240
Downward sloping angle γ/°	12.81
Turns	3.25

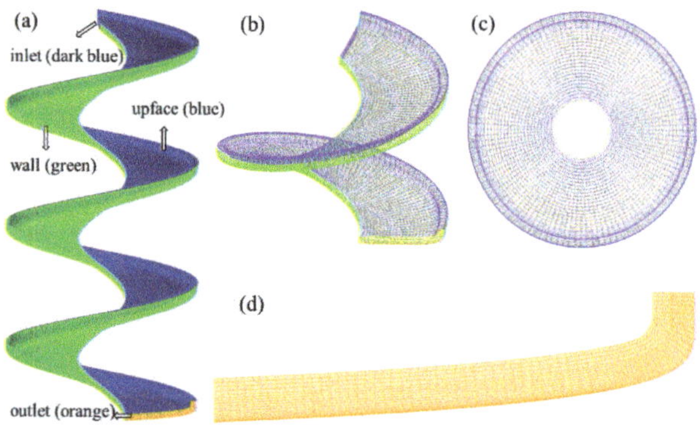

Figure 2. Mesh division of the computational domain of the spiral concentrator. (**a**) Overall computational domain mesh; (**b**) enlarged view of the 1.25-turn computational domain mesh; (**c**) enlarged top view of the computational domain mesh; and (**d**) mesh at the cross-section.

2.2. Numerical Model and Test Conditions

2.2.1. Multiphase Flow Model

Simulation of the gas–liquid–solid multiphase flow is required to capture the changes in the shape of the flow film and the particle motion characteristics in the slurry system. Relevant literature has shown that coupling the VOF (Volume of Fluid) model with the TFM (Two-Fluid Model) can exhibit good prediction performance for this complex system [30–34]. This coupled model is known as the Eulerian Multi-fluid VOF multiphase flow model, which identifies stratified and mixed flows based on the local interfacial diffusion values in the mesh [35]. In systems where stratified and mixed flows coexist, VOF and TFM models are used to capture the interphase interfaces and predict the motion behaviour of discrete-phase particles, respectively [36,37]. Gao et al. have tried to simulate the gas–liquid–solid multiphase flow in the spiral concentrator using the Eulerian Multi-fluid VOF multiphase flow model, and the results of the separation test and numerical simulation show that they can predict well the separation process and indexes of the spiral concentrator under the high-concentration feed ore [26,27,38].

The Eulerian Multi-fluid VOF multiphase flow model is coupled with the VOF model and the TFM model. The governing equations of the VOF model are shown in Equations (1) and (2):

$$\frac{\partial \alpha}{\partial t} + \frac{\partial \alpha}{\partial x_i} u_j = 0 \quad (1)$$

$$\frac{\partial \rho u_i}{\partial t} + \frac{\partial}{\partial x_j}\left(\rho u_i u_j\right) = -\frac{\partial p}{\partial x_i} + \frac{\partial}{\partial x_j}\left[\mu\left(\frac{\partial u_i}{\partial x_j} + \frac{\partial u_j}{\partial x_i}\right)\right] + \rho g_i \quad (2)$$

where α is the volume fraction of a phase (gas or liquid), g_i is the gravitational acceleration and ρ and μ are the average density and viscosity of the fluid, respectively, and are calculated using the following equation:

$$\rho = \alpha\rho_{air} + (1-\alpha)\rho_{water} \quad (3)$$

$$\mu = \alpha\mu_{air} + (1-\alpha)\mu_{water} \quad (4)$$

In the TFM model, each phase has a separate set of control differential equations, and the interphase interactions can be calculated from mass and momentum transfer and coupled phase volume fractions. For this q-phase fluid, the continuity and momentum equations can be expressed as [39]:

$$\frac{\partial}{\partial t}(\alpha_q\rho_q) + \frac{\partial}{\partial x_i}(\alpha_q\rho_q u_{q,j}) = 0 \quad (5)$$

$$\frac{\partial}{\partial t}(\alpha_q\rho_q u_{q,i}) + \frac{\partial}{\partial x_j}\left(\alpha_q\rho_q u_{q,i} u_{q,j}\right) = -\alpha_q\frac{\partial p}{\partial x_i} + \frac{\partial \overline{\overline{\tau}}_{q,ij}}{\partial x_i}\alpha_q\rho_q g_i + \sum\nolimits_{p=1}^{N} R_{pq,i} + F_i \quad (6)$$

where q represents the different phases. As q is taken as 1, it represents the aqueous phase. As q takes 2, it represents the gas phase. As q takes 3 to N, it represents different particulate phases. $\overline{\overline{\tau}}_{q,ij}$ is the stress–strain tensor of phase q. $R_{pq,i}$ is the interphase force term. F_i is the additional force term. $\overline{\overline{\tau}}_{q,ij}$ and $R_{pq,i}$ are expressed as:

$$\overline{\overline{\tau}}_{q,ij} = -\frac{\partial p_q}{\partial x_i}\delta_{ij}\alpha_q\mu_q\left(\frac{\partial u_i}{\partial x_j} + \frac{\partial u_j}{\partial x_i}\right) + \alpha_q\left(\lambda_q - \frac{2}{3}\mu_q\right)\frac{\partial u_{q,i}}{\partial x_i}\delta_{ij} \quad (7)$$

$$R_{pq,i} = K_{pq}(u_{p,i} - u_{q,i}) \quad (8)$$

where p_q is the solid-phase pressure generated by the particle phase, which needs to be considered only in the particle-phase momentum equation. μ_q is the shear viscosity of the fluid; λ_q is the bulk viscosity of the fluid. δ_{ij} is the Kronecker Delta sign. K_{pq} is the interphase momentum exchange coefficient, which is related to the interphase trailing force model.

Fluid properties such as collision behaviour, pressure and shear stress between discrete phase particles can be described using the KTGF model [40]. The momentum exchange coefficient K_{pq} between the liquid and solid phases is described using the Gidaspow model [41]. The momentum exchange coefficient K_{pq} between the liquid and gas phase is described with the Symmetric model [35]. Meng et al. [38] found, in numerical simulation and validation experiments, that numerically simulated iron grade and particle concentration values were low at the inner edge of the spiral concentrator and high in the middle zone. It was determined that the simulation inaccuracy in the existing model could be related to neglect of the Bagnold effect. This prevents the mixed quartz in the lower layer of the flow film from being effectively lifted in the separation process; it then gradually migrates inward under the driving force of the inward flow. When the Bagnold lifting model is adopted, the vertical layering between quartz and hematite particles can be better predicted, which provides favourable conditions for the subsequent radial segregation.

2.2.2. Turbulence Model

The flow on the trough surface of the spiral concentrator involves multiple states of laminar, turbulent and transitional flow. Considering the effect of the low Reynolds number, the RNG (Renormalization Group) k-ε differential formulation theory provides a way to obtain effective viscosity from the analysis. Therefore, the RNG k-ε turbulence model is used in this study to describe the turbulence characteristics of the fluid in the spiral concentrator. The equations of differential transport for turbulent kinetic energy and turbulent dissipation rate are shown below, respectively [42]:

$$\frac{\partial}{\partial t}(\rho k) + \frac{\partial}{\partial x_i}(\rho k u_i) = \frac{\partial}{\partial x_j}\left(\alpha_k \mu_{eff} \frac{\partial k}{x_j}\right) + G_k + G_b - \rho\varepsilon - Y_M + S_k \tag{9}$$

$$\frac{\partial}{\partial t}(\rho\varepsilon) + \frac{\partial}{\partial x_i}(\rho\varepsilon u_i) = \frac{\partial}{\partial x_j}\left(\alpha_k \mu_{eff} \frac{\partial \varepsilon}{x_j}\right) + C_{1\varepsilon}\frac{\varepsilon}{k}(G_k + C_{3\varepsilon}G_b) - C_{2\varepsilon}\rho\frac{\varepsilon^2}{k} - R_\varepsilon + S_\varepsilon \tag{10}$$

where G_k represents the turbulent kinetic energy due to the mean velocity gradient. G_b represents the turbulent kinetic energy due to buoyancy. Y_M represents the fluctuations due to diffusion of the transition in compressible turbulence. μ_{eff} is the effective viscosity (sum of the molecular viscosity and the turbulent viscosity). S_k and S_ε are customized source terms. R_ε is an additional term in the turbulent kinetic energy dissipation rate equation.

2.2.3. Wall Roughness Model

Standard wall functions are used for film flow characteristics within the trough surface. Experiments on rough pipes and channels show that the average velocities near rough walls have the same slope ($1/\kappa$) but different intercepts when plotted on a semilogarithmic scale. The mean velocity near the wall is corrected for roughness as shown in Equation (11) [43]

$$\frac{u_p u^*}{\tau_w/\rho} = \frac{1}{\kappa}\ln\left(E\frac{\rho u^* y_p}{\mu}\right) - \Delta B \tag{11}$$

where u_p is the velocity in the parallel wall direction at a distance y_p from the wall. u^* is the shear flow velocity. τ_w is the wall shear stress. κ is the von Karmen constant. E is an empirical coefficient. y_p is the distance from the wall at point p. μ is the kinetic viscosity. ΔB is a function of the shape and size of the wall roughness.

For sand roughness and similarly uniform roughness elements, ΔB is closely related to the roughness height K_s^+ ($K_s^+ = \rho K_s u^*/\mu$, K_s represents the sand roughness height or diameter). The roughness function is not a single function of K_s^+ but takes different forms depending on the K_s^+ differences. The formula proposed by Bradshaw [44] is usually used to calculate ΔB as shown in Equation (12):

$$\Delta B = \begin{cases} 0 & , \quad K_s^+ \le 2.25 \\ \frac{1}{\kappa}\ln\left(\frac{K_s^+ - 2.25}{87.75} + C_s K_s^+\right) \times sin\{0.4258(\ln K_s^+ - 0.811)\} & , \quad 2.25 < K_s^+ \le 90 \\ \frac{1}{\kappa}\ln[1 + C_s K_s^+] & , \quad K_s^+ > 90 \end{cases} \tag{12}$$

where C_s is the correction factor for the type of roughness, and uniform sand takes 0.5.

The correction of the mean velocity near the wall shifts the logarithmic velocity profile downward, leading to singularities with large K_s and small y^+ values. Therefore, this problem was avoided by moving the wall roughness to half the height of the roughness element while ensuring that the fine mesh was handled correctly [45]. The y^+ value correction at the centre of the first cell is shown in Equation (13).

$$y^+ = y^+ + K_s^+/2 \tag{13}$$

Normally, the numerical equivalent sand roughness height, K_s, is 5 to 10 times the surface roughness, *Ra* [25]. Therefore, the equivalent wall roughness simulation experiment set five K_s values of 0.01, 0.02, 0.03, 0.04 and 0.05 mm. The spiral concentrator separation test system is shown in Figure 3. A flow valve was used to control the return flow to facilitate flow control. In addition, a removable feed device was used to allow interception of the product at the end of the third turn. The spiral concentrator separation experiment used narrow particle size hematite and quartz simple minerals. The purity of the samples was 97% and 96%, respectively. Median particle size was used instead of narrow-grain feed, where the hematite median size was D_{50} = 105.49 μm, and the quartz median size was D_{50} = 107.97 μm. The two particles were prepared as an artificially mixed ore in the ratio of 1:1 by volume (1.87:1.00 by mass). The iron grade in the feed ore was 46.12%, the feed flow rate was 7.2 m^3/h (12 L/min) and the solid mass fraction of the feed ore was 16.74%.

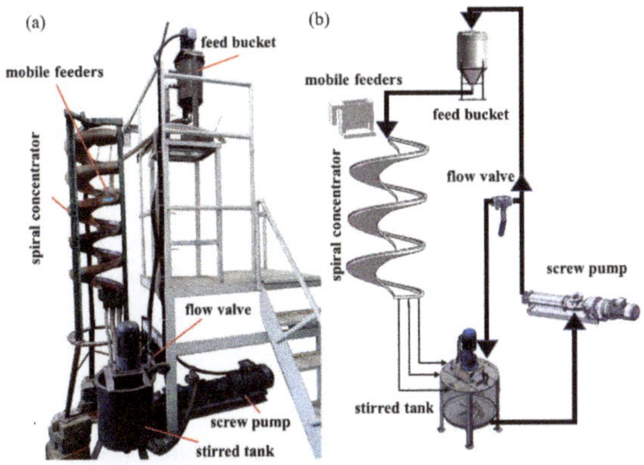

Figure 3. Separation test system of spiral concentrator. (**a**) Experimental system and (**b**) illustration of the separation system.

In the simulation experiment of equivalent wall roughness, the outlet of the spiral concentrator was divided into three product zones: zone I (50~100 mm), zone II (100~165 mm) and zone III (165~200 mm). A fixed container was used to collect the slurry from each zone for 10 s, record the volume of the respective slurry and calculate the slurry flow rate (L/min). The slurry was then dried, weighed, sampled and assayed for iron grade. To ensure the accuracy of the separation experiments, the three experiments were repeated, and the final results were averaged. The comparative information for the slurry flow rate and iron grade of each product zone in the physical and numerical experiments is shown in Figure 4.

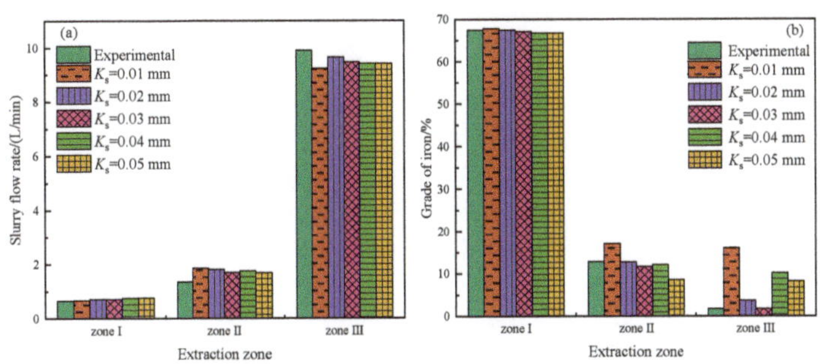

Figure 4. Comparison of experimental and simulated values of product indexes at different wall roughness. (**a**) Slurry flow rate and (**b**) grade of iron.

As can be seen in Figure 4, the measured values of slurry flow rate and grade of iron for the products of zone I are in high agreement with the simulated values at all values of K_s. In contrast, the measured and simulated values for the products of zone II and zone III have some differences. For the slurry flow in zone II and zone III, the K_s values with smaller deviations are 0.03 mm and 0.05 mm, and 0.02 mm and 0.03 mm, respectively. For the iron grades of zone II and zone III products, the K_s values with small deviations are 0.02 mm, 0.04 mm and 0.03 mm, and 0.03 mm and 0.02 mm, respectively. In comprehensive assessment, when K_s = 0.03 mm is selected in numerical simulation, the simulated values of each index obtained have the highest agreement with the measured values.

Based on the correspondence and the average value of the measured wall roughness of the trough Ra = 4.61 μm, the relation $K_s \approx 6.51\ Ra$ is obtained through calculation. The correspondence equation matches the results of related studies [25,46]. Therefore, it is used as the basis for the wall roughness setting in the subsequent numerical simulation experiments.

2.2.4. Experimental Setup and Conditions

The preliminary flow field containing only water and gas phases is first simulated using the Eulerian Multi-fluid VOF multiphase flow model and the RNG k-ε turbulence model. The particulate phase is injected based on the computational stabilisation of the flow field. The simulations are performed using the Eulerian Multi-fluid VOF model that takes into account the Bagnold effect [47]. This model is a coupled Euler–Euler and VOF model, which can examine discrete phases simultaneously and is suitable for sharp and dispersed interfaces. The interphase forces are considered as drag F_d and lift F_l, where the drag model uses the Gidaspow and Symmetric models to predict liquid–solid and gas–liquid interactions, respectively. The lift model loads the Bagnold lift equation into the lift coefficient module by using a user-defined function (UDF).

The boundary of the computational domain is divided into inlet, outlet, upface and wall. The boundary condition for the inlet is velocity inlet. The outlet uses the pressure outlet boundary condition. The wall uses a no-slip boundary condition. Standard wall functions are used for the near-wall treatment. The upface is set as a free-slip boundary condition.

Hematite and quartz are still used as separation objects in the numerical experiment for the adjustment of wall roughness in the middle zone. According to the properties of the samples from the production site, the densities of both were 4950 kg/m^3 and 2650 kg/m^3, respectively. The feed size composition refers to the feeding conditions of the spiral concentrator in the concentrating section of the Qidashan iron ore processing plant [25]. Two particle sizes of 90 μm and 38 μm are set for both minerals, representing coarse and fine particles in the feed, respectively. The four particles are matched according to the mass ratio of 1.87:1.87:1:1 (hematite of 90 μm; hematite of 38 μm; quartz of 90 μm; and quartz of 38 μm). The grade of iron in the feed ore is 45.59%. The slurry solids mass concentration and slurry flow rate are 20% and 0.72 m^3/h (12 L/min), respectively.

2.2.5. Method of Setting Wall Roughness for the Middle Zone

The actual production and use conditions of the spiral concentrator result in variations in trough surface wall roughness, which is not uniform. At the same time, the coarse low-density particles migrate inward and mix with the high-density particles in the middle zone, thus forming a particle mixing zone in the middle zone of the trough surface. Therefore, to investigate the effect of wall roughness variation in the middle zone on the separation process of the spiral concentrator, the 150 mm wide spiral trough surface was divided equally into three radial zones. They are the inner zone (50~100 mm), the middle zone

(100~150 mm) and the outer zone (150~200 mm). The division of radial zones within the single-turn trough surface is shown schematically in Figure 5.

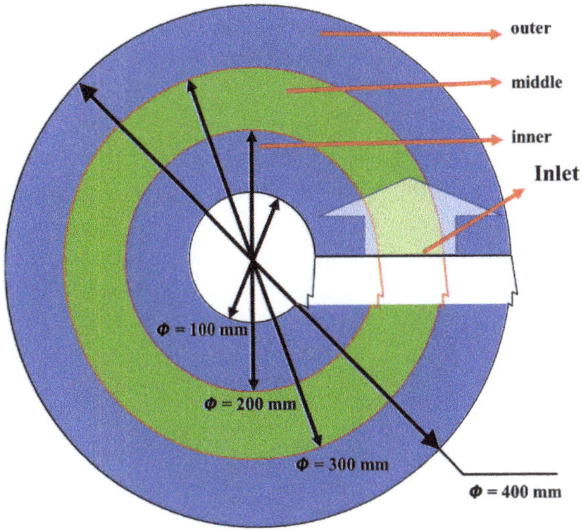

Figure 5. Radial zone divisions of the single-turn trough of the spiral concentrator.

2.3. Data Processing and Methodology

The secondary flow distribution cloud maps are first extracted at intervals of 0.5 turns to facilitate the extraction and analysis of the hydrodynamic parameters. After that, the secondary circulation is divided into inward flow (radial velocity is negative) and outward flow (radial velocity is positive), according to the velocity direction. The area-averaged radial velocities of the fluid are extracted at intervals of 0.25 turns for the three equally divided zones of the inner, middle and outer zones, using which the strength of the secondary flow is characterised.

The evolution of the radial migration behaviour of the fluid and particles is characterised by the variation of the radial flux of the fluid and the cumulative radial flux of the particles with the number of overflow turns, respectively. The fluid radial flux is the radial flow rate of the fluid in the spiral between two adjacent overflow turns at a given radial position. The cumulative radial flux of particles is the cumulative value of the radial flux of particles between the inlet and the target number of overflow turns at a given radial position. The thin-flow film area at the inner and the middle zones is considered to be the main zone of particle accumulation and separation. Five representative radial positions of 60, 80, 100, 120 and 140 mm are selected at equal intervals from the inner zone to the middle zone (50~150 mm) to facilitate the comparison. The data are extracted and analysed separately with a turn extraction interval of 0.125.

In processing the particle distribution data, the trough surface at the end of the 3rd turn, where the flow field develops to a stable level, is divided into six narrow intervals with a width of 25 mm. These are labelled as radial zones r_1~r_6 from inside to outside. In this way, the distribution of hematite and quartz particles in the narrow intervals is examined in detail. Mass flow rates of hematite and quartz particles are processed under different simulation conditions, and particle yields in different radial intervals are calculated sequentially. On this basis, concentrate products with different distribution widths are obtained by adjusting the splitter position. The concentrate yield, iron grade and iron recovery are calculated accordingly. Meanwhile, the separation efficiency of hematite and quartz particles with different particle sizes is calculated using Equation (14) [48]

$$E_s = R_v - R_g \tag{14}$$

where E_s is the separation efficiency, R_v is the recovery of hematite in concentrate, and R_g is the recovery of quartz in concentrate.

3. Results and Discussion

3.1. Effects of Wall Roughness in the Middle Zone on the Distribution and Evolution of Flow Field Parameters in Hematite–Quartz Slurry

Based on the simulation experiment of the equivalent wall roughness mentioned above, the equivalent wall roughness K_s value for the spiral concentrator is determined to be 0.03 mm. Therefore, the overall wall roughness K_s value of 0.03 mm is taken as the base wall. The wall roughness K_s values in the middle zone are set to 0.01, 0.05, 0.1 and 0.2 mm for numerical simulations. The depth of spreading of the slurry and the evolutionary characteristics of the spatial distribution, average velocity and radial flux of the secondary flow are systematically investigated.

3.1.1. Variation of Slurry Spreading Depth

The depth of the slurry flow determines the size of the separation space. Fluid distribution characteristics along the trough surface and the evolution law are especially important for the fluid film concentrator. Figure 6 shows the radial distribution curve of the depth of slurry flow in each turn of the trough surface as the roughness of the middle zone changes.

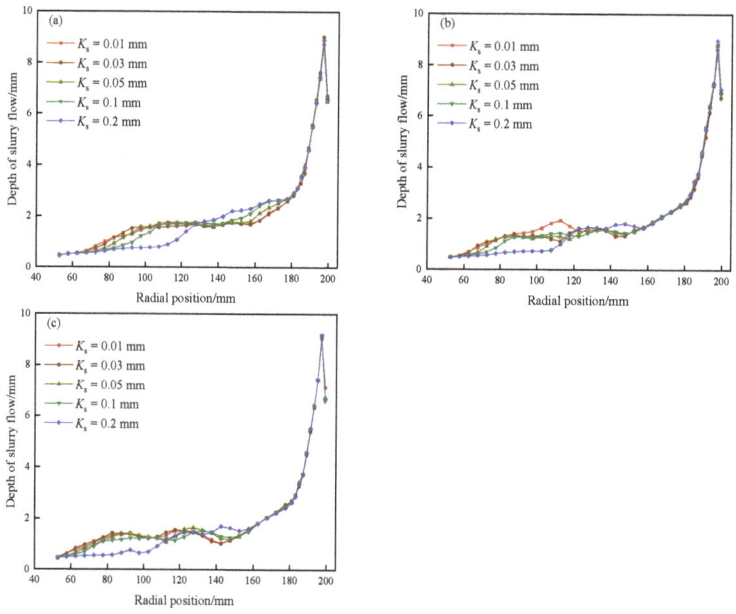

Figure 6. Depth of slurry flow with the adjustment of wall roughness in the middle zone. (**a**) The 1st turn; (**b**) the 2nd turn; and (**c**) the 3rd turn.

As illustrated in Figure 6, the radial distribution characteristics of the slurry flow along the trough surface demonstrate a shallow flow layer at the inner and middle zones and a deep flow layer at the outer zone. As the number of turns increases, the slurry flow gradually spreads inward, and the depth of the slurry flow in the inner edge zone increases.

With the increase of wall roughness, the depth of slurry flow varies in different degrees at different stages. At the end of the first turn, the depth of slurry flow in the inner and middle zones decreases with an increase in wall roughness. The depth of slurry flow increases in the range near the outer edge (130~180 mm). At the end of the second turn, the decrease in the depth of slurry flow at the inner and middle zones slows down. The depth of slurry flow at the outer edge tends to remain constant. In the stable third turn, the slurry flow depth in the inner and middle zones exhibits a slight decrease when the wall roughness K_S value in the middle zone is increased within the range of 0.01~0.1 mm. Furthermore, an increase in wall roughness from 0.1 mm to 0.2 mm leads to a decrease in the depth of slurry flow in the inner edge zone to less than 1 mm, thereby significantly reducing the space available for sorting fluid. This suggests that increasing wall roughness in the middle zone effectively prevents the flow layer from spreading inward, reducing the distribution of slurry flow at the inner edge.

3.1.2. Evolution of the Spatial Distribution of Secondary Flow in the Slurry

Secondary flow constitutes an important form of fluid flow in a spiral concentrator, influencing the spreading behaviour of the flow film along the surface of the trough. Concurrently, it is also a pivotal flow field feature, able to expand the difference in the movement of particles of different densities. Whether the upper and lower flow layers produce relative motion in the radial direction is often the basis for assessing the existence of secondary flow. The spatial distribution of secondary flow in the trough surface with different numbers of turns when adjusting the wall roughness in the middle zone is shown in Figure 7.

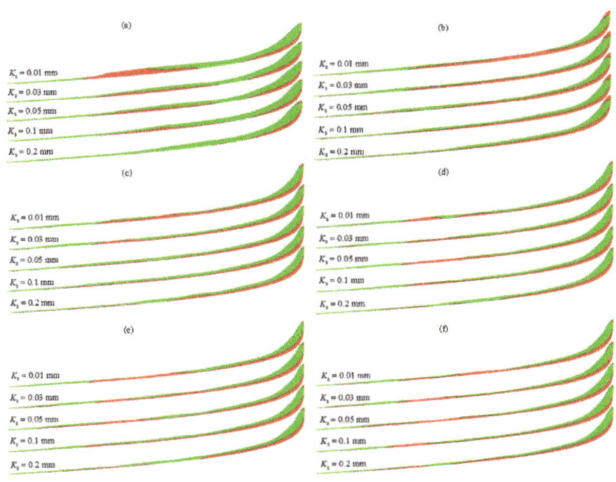

Figure 7. Spatial distribution cloud of secondary flow in the trough at the end of different turns with wall roughness adjustment in the middle zone. (**a**) 0.5 turn; (**b**) 1 turn; (**c**) 1.5 turn; (**d**) 2 turns; (**e**) 2.5 turns; (**f**) 3 turns.

As demonstrated in Figure 7, the distribution area of secondary flow exhibits significant alterations under varying conditions of wall roughness, concomitant with an increase in the number of overflow turns.

(1) In the initial stage (0.5 turn), when the slurry flow is fed into the trough surface, both the inward and outward flow show an obvious intermittent distribution.
(2) Within 1 to 2 turns of the stroke, the inward and outward flow gradually evolved to exhibit continuous distribution characteristics.

(3) Within 2.5 to 3 turns stroke, the upper and lower distributions of the inward and outward flow tend to be stable and consistent.

The overall thickness of the flow layer at the bottom of the trough of the inward flow at the inner and the middle zones decreases gradually with the increase in the number of overflow turns. Increasing the wall roughness in the middle zone, the distribution space of the inward flow significantly reduced in the initial stage. The overall thickness of the flow layer of the inward flow at the inner and the middle zones of the trough surface becomes smaller. The distribution range of the outward flow in the upper layer of the flow film is enlarged. In addition, when the value of wall roughness K_S is increased to 0.2 mm, the inward flow in the middle zone shows an obvious large-scale discontinuity.

3.1.3. Variation of Velocity Distribution of Secondary Flow in the Slurry

The magnitude of the secondary flow velocity is a key parameter that affects the strength of the particle motion migration behaviour. The distribution of the average velocities of the inward flow and the outward flow along the longitudinal stroke (trough length) when adjusting the wall roughness in the middle zone is shown in Figure 8.

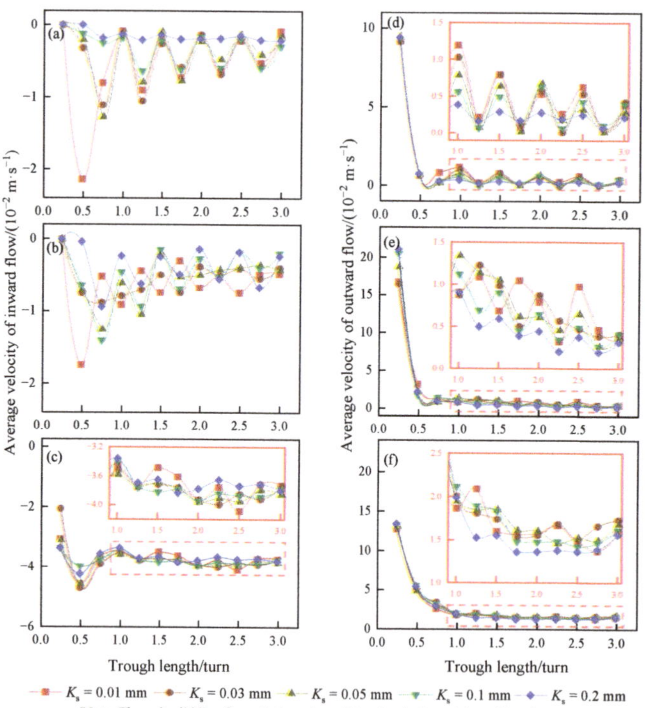

Figure 8. Evolution of the average radial velocity of secondary flow along the longitudinal stroke with wall roughness adjustment in the middle zone. (**a**) Inward flow–inner zone; (**b**) outward flow–middle zone; (**c**) inward flow–outer zone; (**d**) outward flow–inner zone; (**e**) outward flow–middle zone; and (**f**) outward flow–outer zone.

As shown in Figure 8, the secondary flow means velocity in all zones fluctuates in different wall roughness conditions. The fluctuations in the secondary flow mean velocity decreases with an increase in the longitudinal stroke.

As illustrated in Figure 8, the average velocity of inward flow in the inner edge zone approaches 0 m/s. It remains relatively stable throughout the longitudinal stroke when the wall roughness K_s is set to 0.2 mm. This phenomenon is associated with decreased slurry flow distribution at the inner zone, as previously discussed. Conversely, when the remaining wall roughness is considered, the average velocity of inward flow in the inner and middle zones varies significantly within 1.5 turns, subsequently demonstrating a fluctuating decline. The mean velocity of inward flow in the outer zone exhibits a more uniform trend across all wall roughness conditions, increasing and then decreasing within the first turn, followed by slight fluctuations. Notably, the average velocity of the inward flow in the inner and middle zones stabilises at approximately 2.5~3 turns. As the wall roughness increases in the middle zone, there is a tendency for the average velocity of the inward flow at the inner and middle zones to decrease. When the value of K_s is increased to 0.2 mm, the mean velocity of the inward flow in both zones decreases significantly. In contrast, the average velocity of the inward flow in the outer zone decreases only in the first turn of the stroke, with subsequent changes being random.

The mean velocity of the outward flow in all zones exhibits a similar trend in the longitudinal stroke. A notable decrease occurs within 0.5 turns of the stroke, followed by a gradual decline. Increasing the wall roughness in the middle zone, the average velocity of outward flow in all zones showed a decreasing trend from 2.5 to 3 turns of stroke. The decreasing trend is more significant in the inner and the middle zones.

3.1.4. Distribution and Evolution of Radial Fluxes of Secondary Flow

The evolutionary variation of the secondary flow radial flux on the longitudinal stroke can reflect the radial migration characteristics of the secondary flow and its stability. The previous analyses show that the average velocities of the slurry flow and secondary flow vary significantly in the inner and middle zones. This is the zone of enrichment for high-density particles. Hence, further focus is given to this zone. The evolution of radial fluxes of inward flow and outward flow along the longitudinal stroke (trough length) for different wall roughness of the middle zone is shown in Figure 9.

As presented in Figure 9, the secondary flow radial flux along the longitudinal stroke shows comparable trends for varying radial positions. The peak radial fluxes manifested within the first half of a turn, subsequently exhibiting fluctuating variations and stabilising within a certain range in the stroke. Specifically, the peak radial flux of the outward flow is considerably larger than the peak radial flux of the inward flow. The radial fluxes of the inward and outward flows are essentially close to being in the steady state stage. This is consistent with the net radial flux of the fluid being zero in the steady state limit, as derived by Lee et al. [49].

Further from Figure 9, it can be observed that the radial fluxes of both inward flow and outward flow in the inner and middle zones decrease to different degrees with increasing wall roughness of the middle zone. The radial flux of the secondary flow exhibits a negative correlation with the magnitude of the wall roughness values at 60 mm and 80 mm in the inner zone. Compared with the inner zone, the fluctuation of the radial flux of the fluid in the middle zone is increased. However, the fluctuation of the radial flux of the secondary flow in the middle zone tends to decrease with increasing wall roughness. As illustrated in Figure 9, the radial flux of the outward flow tends to decrease in the inner and middle zones, with increasing wall roughness in the middle zone.

Analyses for the secondary flow radial fluxes demonstrate that an increase in wall roughness (K_s value) in the middle zone within the range from 0.01 to 0.1 mm results in a substantial reduction of the radial fluxes of the inward and outward flows in the inner and middle zones of the trough surface. The inward and outward flow of the fluid will be reduced. However, the radial flux of the inward and outward flow of the fluid tends to zero in the inner and middle zones of the trough surface when the value of K_s is increased in the range from 0.1 to 0.2 mm. It indicates that the fluid radial flow is weak in this roughness range. This finding is consistent with the observed trend of spreading motion of the flow film. Changes in the strength of the radial flow trend of the secondary flow will inevitably impact the migration motion and spatial distribution of solid particles.

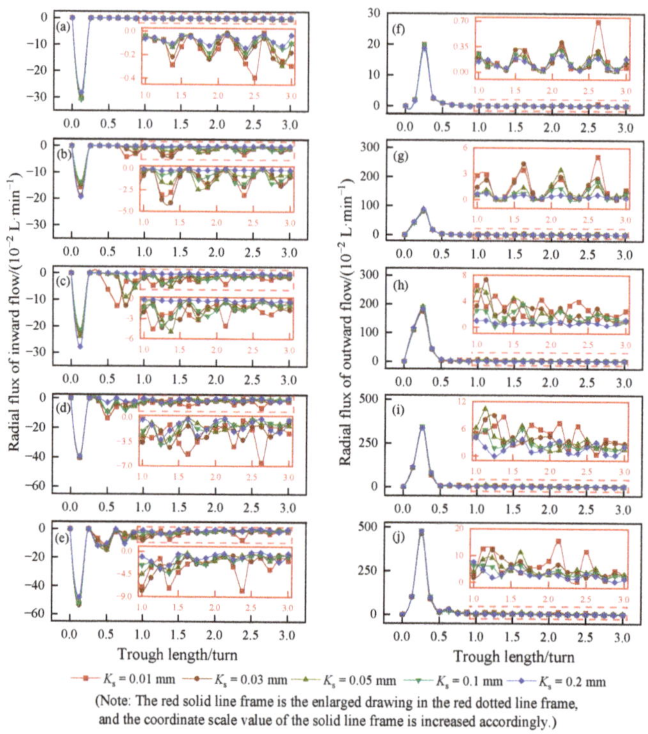

Figure 9. Evolution of radial flux of secondary flow along the longitudinal stroke with adjustment of wall roughness in the middle zone. (**a**) Inward flow–r = 60 mm; (**b**) inward flow–r = 80 mm; (**c**) inward flow–r = 100 mm; (**d**) inward flow–r = 120 mm; (**e**) inward flow–r = 140 mm; (**f**) outward flow–r = 60 mm; (**g**) outward flow–r = 80 mm; (**h**) outward flow–r = 100 mm; (**i**) outward flow–r = 120 mm; and (**j**) outward flow–r = 140 mm.

3.2. Influence of Wall Roughness in the Middle Zone on Particle Migration Motions and Separation Indexes

3.2.1. Variation in the Radial Migration Behaviour of Hematite and Quartz Particles

The cumulative radial fluxes of hematite and quartz particles in the inner and middle zones of the trough surface are processed. This is used to analyse the along-stroke variation of the radial migration behaviour of the particles and the effect of the wall roughness of the middle zone on this. The results are shown in Figure 10.

From Figure 10, it can be found that the cumulative radial flux of hematite particles is negative at r = 60 mm. This indicates that the hematite particles are migrating inward with little migration. In the remaining four radial positions, the hematite particles migrate outward and then inward as the longitudinal stroke of the fluid is prolonged, and the amount of migration is large. As the radial position shifts outward, the cumulative radial flux of hematite tends to stabilise in a shorter stroke. Coarse-grained hematite reaches migration equilibrium more quickly than fine-grained hematite. At r = 60 mm at the inner edge, the inward migration of hematite shows a decreasing trend with increasing wall roughness of the middle zone only in 2~3 turns of the stroke. At r = 80 mm and r = 100 mm, the inward migration of hematite decreases significantly with increasing wall roughness of the middle zone. At r = 120 mm and r = 140 mm, the inward migration of hematite is also negatively correlated with the wall roughness of the middle zone. The hematite particles all reached migration equilibrium between 2 and 3 turns of the stroke. When K_s varies in the range from 0.01 to 0.1 mm, the difference in hematite particle migration is insignificant. When the value of K_s increased from 0.1 to 0.2 mm, the inward migration of hematite particles decreased significantly.

As shown in Figure 10, the cumulative radial fluxes of quartz at the inner edge r = 60 mm are also both negative. This indicates inward migration and a small amount of migration. The quartz particles migrated outward at the other radial positions, and the cumulative radial flux increased with the radial position outward. However, at the radial position r = 140 mm, the coarse-grained quartz gradually migrates inward after roughly 1.5 turns. Increasing the wall roughness of the middle zone resulted in no significant change in the cumulative radial flux of quartz at the inner edge at r = 60 mm. At both r = 80 mm and r = 100 mm, the outward migration of both particle sizes of quartz increased continuously with roughness from over 0.5 to 2 turns of the stroke. The migration difference due to increased roughness decreases in subsequent strokes. At both r = 120 mm and r = 140 mm, the outward migration of quartz of both particle sizes also increases with increasing roughness. The migration difference increases with longer longitudinal strokes.

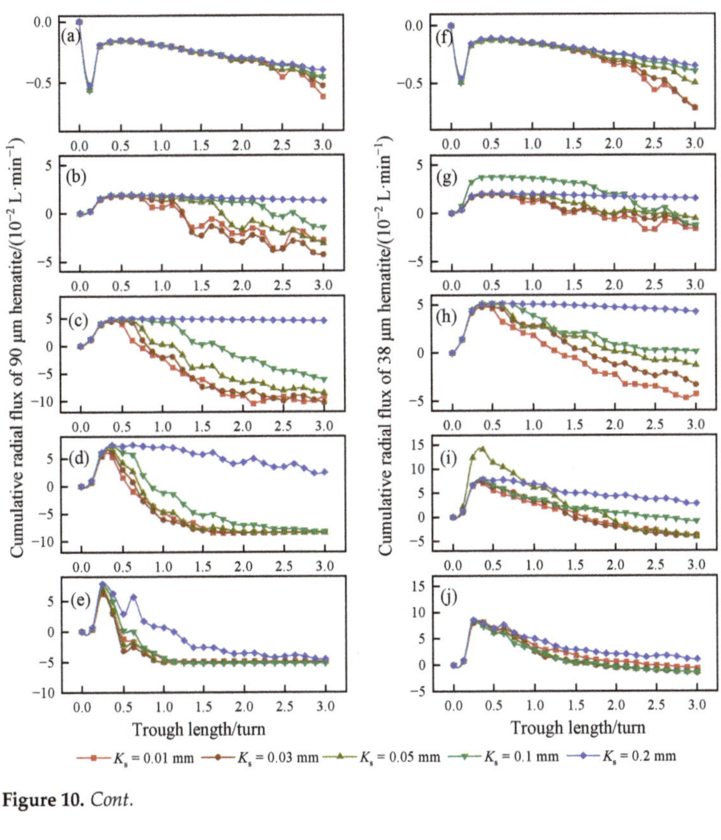

Figure 10. *Cont.*

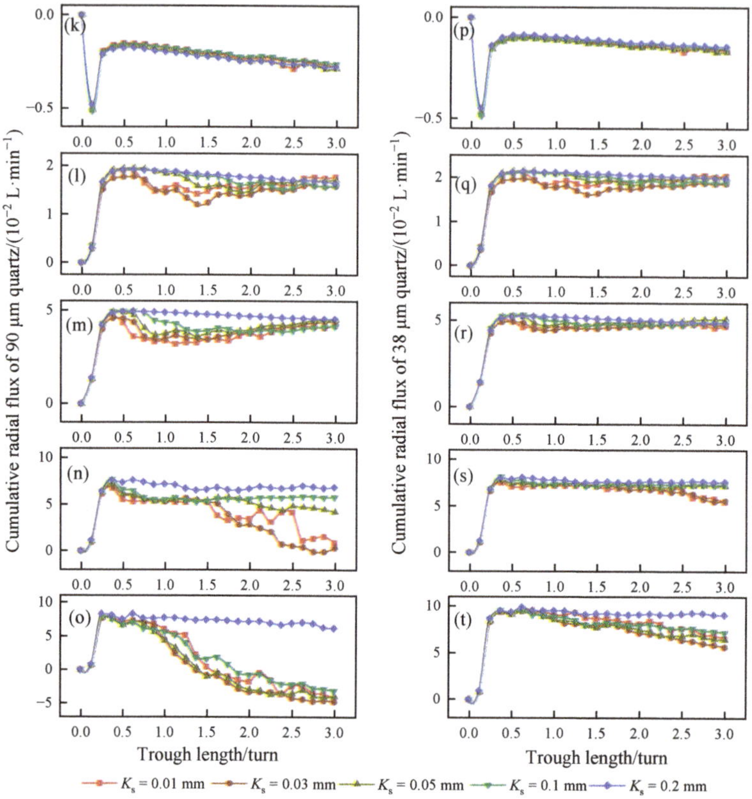

Figure 10. Evolution of particle radial fluxes along the longitudinal stroke with adjustment of wall roughness in the middle zone. (**a**) 90 μm hematite–r = 60 mm; (**b**) 90 μm hematite–r = 80 mm; (**c**) 90 μm hematite–r = 100 mm; (**d**) 90 μm hematite–r = 120 mm; (**e**) 90 μm hematite–r = 140 mm; (**f**) 38 μm hematite–r = 60 mm; (**g**) 38 μm hematite–r = 80 mm; (**h**) 38 μm hematite–r = 100 mm; (**i**) 38 μm hematite–r = 120 mm; (**j**) 38 μm hematite–r = 140 mm; (**k**) 90 μm quartz–r = 60 mm; (**l**) 90 μm quartz–r = 80 mm; (**m**) 90 μm quartz–r = 100 mm; (**n**) 90 μm quartz–r = 120 mm; (**o**) 90 μm quartz–r = 140 mm; (**p**) 38 μm quartz–r = 60 mm; (**q**) 38 μm quartz–r = 80 mm; (**r**) 38 μm quartz–r = 100 mm; (**s**) 38 μm quartz–r = 120 mm; (**t**) 38 μm quartz–r = 140 mm.

The analyses demonstrate that an augmentation in the wall roughness of the middle zone results in a diminution of the inward migration of hematite from the inner and middle zones. Concurrently, it engenders an increase in the outward migration of quartz particles. Moreover, the migration difference between hematite and quartz is substantial. Subsequent analysis of the difference in the distribution of particles within the trough surface and the separation index is requisite to investigate the separation behaviour.

3.2.2. Spatial Distribution Characteristics of Hematite and Quartz Particles

Based on the evolution of the flow field in a spiral concentrator and the equilibrium characteristics of particle motion, the yields of four types of particles in six radial zones in the trough surface at the end of the third turn are further processed. The distribution characteristics of different particles in the trough surface and their changes with the adjustment of the wall roughness of the middle zone are systematically investigated. The experimental results are shown in Figure 11.

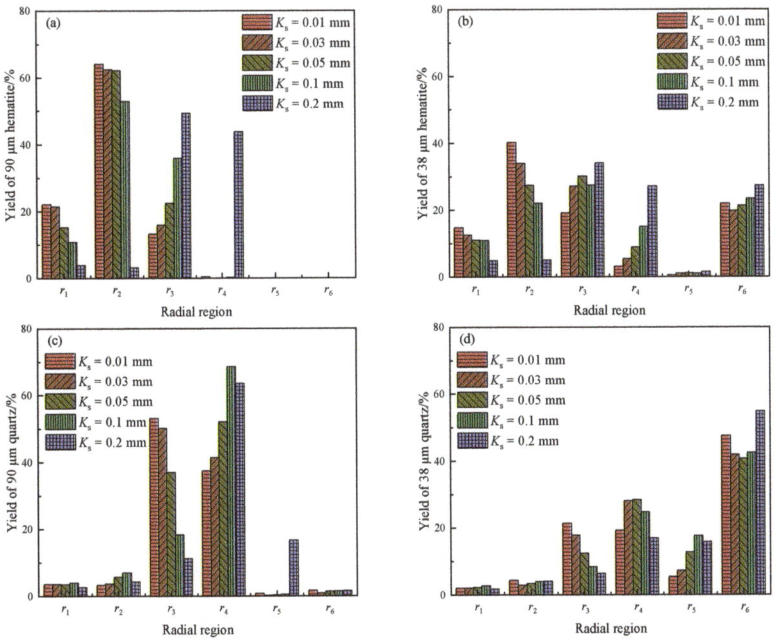

Figure 11. Distribution of particle yield in different radial zones with adjustment of wall roughness in the middle zone. (**a**) 90 μm hematite; (**b**) 38 μm hematite; (**c**) 90 μm quartz; and (**d**) 38 μm quartz.

From Figure 11, it can be seen that the hematite particles are mainly distributed in the r_1~r_3 zone (the inner half of the trough). The main distribution areas of quartz particles are r_3, r_4 and the outer zone. The distribution of 38 μm hematite is also higher in the outer half of the trough due to the fine particle flow following effect. The total yield of hematite in the inner half of the trough is 99.58% and 74.25% for both particle sizes at K_s = 0.01 mm for the wall roughness of the middle zone. The yield of 90 μm hematite is 25% higher than that of 38 μm hematite. The 38 μm hematite yielded 22.01% in the r_6 zone at the outermost edge. It can be seen that 38 μm hematite is highly susceptible to loss in the outer zone. The total yield of both quartz sizes in the middle r_3 and r_4 zones is 90.50% and 40.58%, respectively. The 38 μm quartz is nearly 50% lower than the distribution of 90 μm quartz in the middle zone. The yield of 38 μm quartz in the outermost r6 zone can reach 47.59%. This indicates that most of the fine quartz particles are excluded from the outer zone in the separation process. It can be seen that the middle zone of the trough surface is the mixed zone of coarse quartz and hematite particles, which has an important effect on the particle separation results.

As seen in Figure 11a,b, increasing the wall roughness of the middle zone significantly reduces the hematite yield in the r_1 and r_2 zones (inner zone). As the wall roughness increases from 0.01 mm to 0.1 mm, the yield of coarse and fine hematite particles in the inner zone decreases from 86.32% and 55.02% to 63.80% and 33.06%, respectively. For K_s = 0.2 mm, it decreases sharply to 6.97% and 9.87%, respectively. The hematite yield in the middle zone increased accordingly, particularly in the mixed zone r_3, where the increase in hematite yield was significant. This suggests that increasing the wall roughness of the middle zone reduces the migration of hematite particles to the concentrated zone within the trough surface. The overall distribution of hematite is shifted outward. The enrichment of hematite particles in the inner zone decreases sharply if the wall roughness is too high.

As shown in Figure 11c,d, the quartz yield in the r_1 and r_2 zones (inner zone) varies less with the increase in wall roughness. As the wall roughness K_s increases from 0.01 mm to 0.1 mm in the middle zone, the yield changes significantly in the r_3 and r_4 zones in the middle of the trough surface. The 90 μm quartz yield decreases from 53.05% to 18.31% and the 38 μm quartz yield decreases from 21.35% to 8.41% in the r_3 zone. Thus, the quartz distribution in the inner half of the trough decreases significantly with increasing roughness. In the r_4 zone, the 90 μm quartz yield increases from 37.45% to 68.62%. The 38 μm quartz yield increases from 19.23% to 24.82%. The quartz particle yield distribution change is reflected as a change from the r_3 zone to the r_4 zone, with a greater change in the coarse quartz. As K_s increases from 0.1 mm to 0.2 mm, the quartz yield decreases in both the r_3 and r_4 zones, and the total yield increases correspondingly in the outer r_5 and r_6 zones.

In summary, it shows that increasing wall roughness in the middle zone decreases the enrichment of hematite particles in the two micro-zones in the inner zone. The enrichment in the two micro-zones in the middle zone increases. The zone of quartz particle enrichment is shifted outward. The distribution and change trend of the mineral particles in the trough surface show obvious differences in density and particle size. This is consistent with the previously described trend of radial particle flux along the stroke.

3.2.3. Variation of Separation Indexes of Hematite and Quartz Particles

The concentrate products are obtained by adjusting the splitter location at the trough of the third turn to visualize the trend of separation indexes with the adjustment of the wall roughness in the middle zone. The corresponding concentrate yield, grade of iron and recovery rate of iron are counted, respectively. The experimental results are shown in Figure 12.

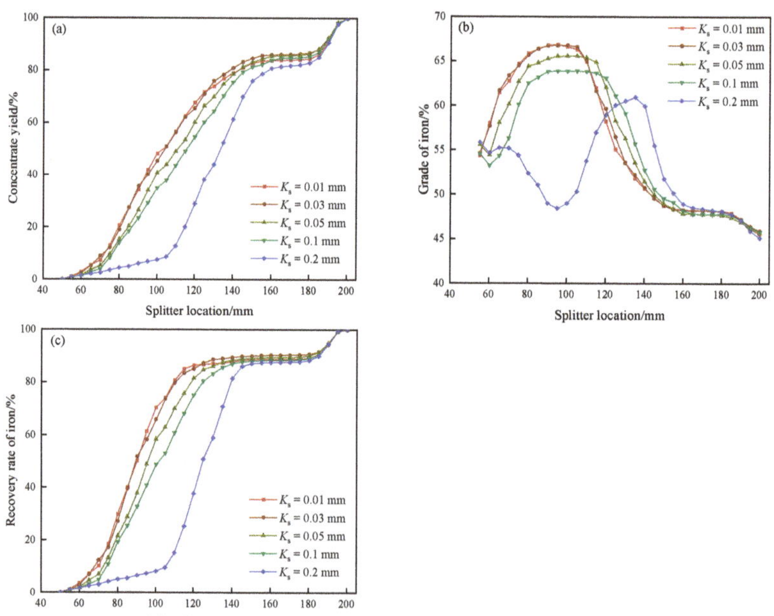

Figure 12. Variation of concentrate indexes with splitter location at different wall roughness in the middle zone. (**a**) Concentrate yield; (**b**) grade of iron; and (**c**) recovery of iron.

It is shown in Figure 12 that the concentrate yield increases overall with the outward shift of the splitter location (product segmentation location) under different wall roughness. Except for the condition of a K_s value of 0.2 mm, the iron grade of concentrate shows a general trend of increasing and then decreasing with the outward shift of the splitter. The highest grade of iron in the product can be obtained by intercepting the ore at the appropriate radial location. The trend in the recovery of iron is similar to the trend in yield. In addition, the highest grade of iron in concentrate is distributed in the range of 60.94%–66.83%. With the increase of wall roughness in the middle zone, the corresponding maximum grade of iron in concentrate decreased, and the location of the splitter shifted outward. Concentrate yield and the recovery of iron decreased.

Specifically, when the K_s values of the wall roughness in the middle zone are 0.01 and 0.03 mm, the difference in the separation indexes of the spiral concentrator is small. When the value of K_s increases from 0.03 mm to 0.1 mm, the separation index decreases within a degree. However, as the K_s value increases from 0.1 mm to 0.2 mm, the maximum grade of iron in the concentrate decreases significantly. The radial distribution curves of both yield and recovery of iron are significantly shifted to the direction outside the trough. This indicates that the separation result is worse. Therefore, to avoid the decrease of the production indexes caused by the high wear of the trough surface, it is necessary to control the central wall roughness K_s value at no more than 0.1 mm (*Ra* test value not more than 15.36 μm). Concentrate indicators and interception conditions at suitable wall roughness can be referred to in the experimental data in Table 2.

Table 2. Concentrate indexes and splitter conditions at suitable wall roughness.

Wall Roughness (K_s) /mm	Surface Roughness (*Ra*) /μm	Maximum Iron Grade /%	Splitter Location /mm	Concentrate Yield /%	Recovery of Iron /%
0.01	1.36	66.83	95	41.86	61.43
0.03	4.61	66.74	100	45.30	65.87
0.05	7.68	65.55	100	40.73	58.42
0.1	15.36	63.86	105	37.95	52.97

3.2.4. Comparison of Separation Efficiency of Hematite and Quartz Particles with Different Sizes

Further calculations are carried out on the separation efficiency of hematite and quartz particles with different sizes at different wall roughness. The effect of wall roughness in the middle zone on the separation efficiency of combining particles with different particle sizes is investigated. The results are shown in Figure 13.

It can be observed from Figure 13 that, as the splitter location is shifted outward, the separation efficiency curves of hematite and quartz particles generally show a peaked distribution. There exists a maximum separation efficiency and an optimal product splitting location. The maximum values of separation efficiencies achieved for the four particle combinations are 83.98%, 85.84%, 52.38% and 55.33%, respectively, when the wall roughness K_s = 0.01 mm in the middle zone. It can be seen that the particle separation efficiency of 90 μm hematite, either in combination with 90 μm quartz or with 38 μm quartz, is significantly higher than that of the separation efficiency of the 38 μm hematite separation system. Both with 90 μm hematite and with 38 μm hematite, a more efficient separation is obtained with 38 μm quartz as the gangue mineral.

As can be seen from Figure 13a,b, the maximum separation efficiency of 90 μm hematite from quartz shows a tendency to increase and then decrease with the increasing value of wall roughness K_s of the middle zone in the range from 0.01 mm to 0.1 mm.

However, it can be maintained at a high level. Both combinations obtain the maximum separation efficiency (84.42%, 89.36%) in the middle zone with a wall roughness K_S = 0.03 mm. As the value of K_S increases from 0.1 mm to 0.2 mm, the maximum separation efficiencies of the two combinations decrease significantly to 67.12% and 78.46%, respectively. The optimum interception location is significantly shifted outward.

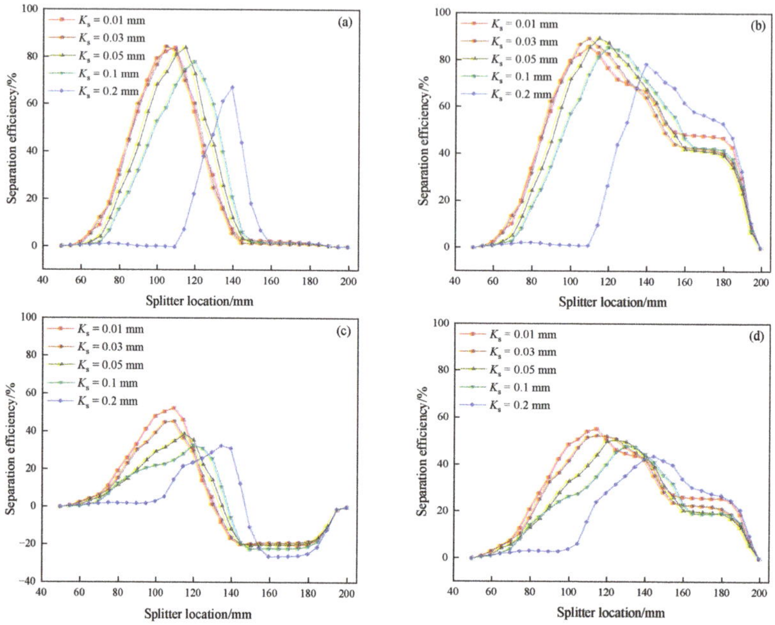

Figure 13. Variation of particle separation efficiency with splitter location at different wall roughness in the middle zone. (**a**) 90 μm hematite + 90 μm quartz; (**b**) 90 μm hematite + 38 μm quartz; (**c**) 38 μm hematite + 90 μm quartz; and (**d**) 38 μm hematite + 38 μm quartz.

From Figure 13c,d, it can be found that the maximum separation efficiency of 38 μm hematite from quartz decreases significantly with the increase of wall roughness in the middle zone, while the optimum interception location shifts outward. When the value of wall roughness K_S = 0.2 mm, the maximum separation efficiency decreases to 32.49% and 43.71%, respectively. Among these, the separation efficiency decreases more significantly for the combination of 38 μm hematite and 90 μm quartz particles. This is closely related to the high distribution of coarser quartz in the middle of the trough surface. Furthermore, the distribution of 38 μm hematite is predominantly concentrated at the outermost edge. Conversely, 90 μm quartz exhibits a notable enrichment within the middle zone. Consequently, when the splitter is located in the outer zone, the separation efficiency of this combination exhibits a negative value.

The preceding data set indicates that the separation efficiency of hematite and quartz is affected by the interplay between the feed particle size and the wall roughness within the middle zone. The initial value of K_S (Ra = 4.61 μm) in the middle zone is deemed appropriate to facilitate the recovery of coarser hematite particles. To facilitate the recovery of fine hematite particles, it is recommended that an initial K_S value of 0.01 (Ra = 1.41 μm) be employed. Within the context of the spiral concentrator processing and fabrication process, it is important to ensure that the trough surface is maintained close to the smooth wall condition. The production process needs to control wear to prevent excessive partial wall roughness. As above, analysis and assessment need to be verified by further actual experiments.

4. Conclusions

In this study, a simulation of rough wall surfaces based on laboratory tests was carried out. By means of numerical simulation, the influence of wall roughness K_s in the middle zone of the spiral trough on the flow field evolution and particle separation behaviour of hematite–quartz–water slurry is studied. The suitable initial value and the control range of the wall roughness suitable for different-sized particles to obtain a better separation index are determined. The main conclusions are as follows.

1. The slurry fluid needs to develop a certain stroke of evolution to form a continuous secondary flow distribution in the spiral concentrator. Increasing the value of wall roughness K_s in the middle zone within the range of 0.01~0.1 mm, the depth of the slurry flow in the inner and the middle zone is slightly reduced. The separation fluid space is severely reduced when the K_s value is increased to 0.2 mm. The inward flow shows obvious discontinuity.
2. The longitudinal evolution of velocities and radial fluxes of the secondary flow exhibits fluctuations and tends to equilibrium gradually. The velocities of inward and outward flow show a decreasing trend with an increase in the wall roughness of the middle zone. The radial flux of the fluid decreases more at the inner and middle zones of the trough, especially when K_s increases from 0.1 mm to 0.2 mm.
3. The inward migration of hematite particles decreases and the outward migration of quartz particles increases as the degree of wall roughness of the middle zone increases, resulting in the outward shifting of particles' enrichment zones. The distribution and variation patterns of the two minerals in the trough exhibit remarkable differences in density and particle size, which is consistent with the trend of the along-track radial flux of the particles.
4. A number of concentrate products can be obtained by adjusting the splitter location in the numerical experiments. The separation index decreases slightly as the wall roughness K_s value in the middle zone increases from 0.01 to 0.1 mm, but the iron grade of concentrate decreases obviously as the K_s value increases to 0.2 mm.
5. The separation effect of hematite and quartz particles in the spiral concentrator is influenced by the comprehensive interaction of feed size and wall roughness in the middle zone. The initial value of wall roughness in the middle zone of the spiral trough should be differentiated to recover hematite particles of different sizes. It is also necessary to reduce wear in production to prevent wall roughness from increasing to a harmful range.

Author Contributions: S.G.: conceptualization, methodology, writing—review and editing, project administration, resources and data curation. X.Z.: writing—original draft, software, data curation, formal analysis and visualisation. B.L.: investigation, software and data curation. Q.W.: investigation, software and validation. C.L.: software and data curation. All authors have read and agreed to the published version of the manuscript.

Funding: This research was funded by the [National Natural Science Foundation of China] grant number [52274257], the [Project of Zhongyuan Critical Metals Laboratory] grant number [GJJSK-FYB202406] and the [National Natural Science Foundation of China] grant number [51974065].

Data Availability Statement: Data will be made available on request.

Acknowledgments: The authors gratefully acknowledge financial support from the National Natural Science Foundation of China (Grant No. 52274257), the Project of Zhongyuan Critical Metals Laboratory (Grant No. GJJSKFYB202406) and the National Natural Science Foundation of China (Grant No. 51974065).

Conflicts of Interest: The authors declare that they have no known competing financial interests or personal relationships that could have appeared to influence the work reported in this paper.

References

1. Jain, P.K.; Rayasam, V. An analytical approach to explain the generation of secondary circulation in spiral concentrators. *Powd Technol.* **2017**, *308*, 165–177. [CrossRef]
2. Jain, P.K. A novel concept of composite coefficient of friction for slurry systems and development of flow equations for spir concentrators. *Powder Technol.* **2021**, *386*, 411–419. [CrossRef]
3. Ye, G.; Huo, Y.; Li, C.; Deng, C.; Yu, Y.; Huang, G.; Ma, L. A comparative study of trough profile and operating paramete performance in spiral concentrator. *Int. J. Coal Prep. Util.* **2021**, *41*, 678–691. [CrossRef]
4. Holland-Batt, A.B. Some design considerations for spiral separators. *Miner. Eng.* **1995**, *8*, 1381–1395. [CrossRef]
5. Atasoy, Y.; Spottiswood, D.J. A study of particle separation in a spiral concentrator. *Miner. Eng.* **1995**, *8*, 1197–1208. [CrossRef]
6. Matthews, B.W.; Fletcher, C.A.J.; Partridge, A.C.; Vasquez, S. Computations of curved free surface water flow on spiral concent tors. *J. Hydraul. Eng.* **1999**, *125*, 1126–1139. [CrossRef]
7. Dixit, P.; Tiwari, R.; Mukherjee, A.K.; Banerjee, P.K. Application of response surface methodology for modeling and optimizatic of spiral separator for processing of iron ore slime. *Powder Technol.* **2015**, *275*, 105–112. [CrossRef]
8. Gao, S.; Meng, L.; Wei, D.; Song, Z.; Yuan, J. Research advance of flow characteristics in spirals and its separation performanc *Conserv. Util. Miner. Resour.* **2020**, *40*, 166–171. (In Chinese) [CrossRef]
9. Romeijn, T.; Behrens, M.; Paul, G.; Wei, D. Experimental analysis of water and slurry flows in gravity-driven helical miner separators. *Powder Technol.* **2022**, *405*, 117538. [CrossRef]
10. Luo, K.; Dai, Q.; Liu, X.; Fan, J. Effects of wall roughness on particle dynamics in a spatially developing turbulent boundary lay *Int. J. Multiph. Flow* **2019**, *111*, 140–157. [CrossRef]
11. Keirsbulck, L.; Labraga, L.; Mazouz, A.; Tournier, C. Surface roughness effects on turbulent boundary layer structures. *J. Flui Eng.* **2002**, *124*, 127–135. [CrossRef]
12. Sun, Z.; Li, X.; Zhao, S.; Zhong, T.; Wang, L.; Yan, X.; Zhang, H. Enhanced selective recovery of fine minerals in flotation by th oscillatory turbulence. *J. Environ. Chem. Eng.* **2025**, *13*, 115164. [CrossRef]
13. Li, X.; Sun, Z.; Wang, H.; Wang, L.; Yan, X.; Zhang, H.; Ran, J.; Liu, J. Particle motion under turbulent eddies: Inspiration for fir minerals flotation. *Chem. Eng. Sci.* **2025**, *301*, 120754. [CrossRef]
14. Wang, H.; Chen, R.; Li, D.; Zhang, B.; Yan, X.; Ran, J.; Zhang, H. Probing the adaptability relationship between flow characteristi and mineral particles with different size and surface hydrophobicity. *Powder Technol.* **2025**, *452*, 120588. [CrossRef]
15. Zhou, R.; Li, X.; Li, D.; Wang, H.; Song, Z.; Yan, X.; Zhang, H. The effect of energy input on bubble-particle collision, attachmen detachment, and collection efficiencies in a mechanical flotation cell. *Powder Technol.* **2025**, *453*, 120659. [CrossRef]
16. Davies, P.O.J.; Goodman, R.H.; Deschamps, J.A. Recent developments in spiral design, construction and application. *Miner. En* **1991**, *4*, 437–456. [CrossRef]
17. Jiménez, J. Turbulent flows over rough walls. *Annu. Rev. Fluid Mech.* **2004**, *36*, 173–196. [CrossRef]
18. Orlandi, P.; Leonardi, S. DNS of turbulent channel flows with two- and three-dimensional roughness. *J. Turbul.* **2006**, *7*, N7 [CrossRef]
19. Bons, J.P.; McClain, S.T.; Wang, Z.J.; Chi, X.; Shih, T.I. A comparison of approximate versus exact geometrical representations c roughness for cfd calculations of cf and St. *J. Turbomach.* **2008**, *130*, 021024. [CrossRef]
20. Prakash, A.; Laurendeau, E. Consistent surface roughness extension for wall functions. *Int. J. Heat Fluid Flow* **2020**, *82*, 10855 [CrossRef]
21. Arnold, D.J.; Stokes, Y.M.; Green, J.E.F. Thin-film flow in helically wound shallow channels of arbitrary cross-sectional shap *Phys. Fluids* **2017**, *29*, 013102. [CrossRef]
22. Romeijn, T.; Fletcher, D.F.; de Andrade, A. Evaluation of numerical approaches for the simulation of water-flow in gravity-drive helical mineral separators. *Sep. Sci. Technol.* **2023**, *58*, 2519–2538. [CrossRef]
23. Zhou, F.; Sun, G.; Zhang, Y.; Ci, H.; Wei, Q. Experimental and CFD study on the effects of surface roughness on cyclon performance. *Sep. Purif. Technol.* **2018**, *193*, 175–183. [CrossRef]
24. Gao, S.; Zhou, X.; Wang, Q.; Liu, W. Evolution characteristic of slurry flow field parameters and particles separation behavior i spirals with rough wall. *Conserv. Util. Miner. Resour.* **2023**, *43*, 127–136. (In Chinese) [CrossRef]
25. Zhao, Q.; Cui, B.; Hou, D.; Wei, D.; Song, T.; Feng, Y. Effects of wall roughness on the separation performance of hydrocyclone under different inlet conditions. *Ind. Eng. Chem. Res.* **2021**, *60*, 11251–11266. [CrossRef]

36. Gao, S.; Meng, L.; Zhou, X.; Shen, Y.; Cui, B.; Song, Z. Design of partial cross-sectional geometry and prediction of separation performance in the spiral separator. *Sep. Sci. Technol.* **2022**, *57*, 2127–2144. [CrossRef]
37. Gao, S.; Zhou, X.; Meng, L.; Zhao, Q.; Liu, W. Variation of flow hydrodynamic parameters and prediction of particle separation indices in the spiral concentrator with the regulation of pitch-diameter ratio. *Separations* **2023**, *10*, 410. [CrossRef]
38. Gao, S.; Wang, Q.; Zhou, X.; Liu, C.; Shen, Y.; Cui, B. Effect of the cross-sectional geometry of the mixed particle zone on the spiral separation process and its structural optimization. *Minerals* **2024**, *14*, 1251. [CrossRef]
39. Meng, L.; Gao, S.; Wei, D.; Cui, B.; Shen, Y.; Song, Z.; Yuan, J. Effects of cross-sectional geometry on flow characteristics in spiral separators. *Sep. Sci. Technol.* **2021**, *56*, 2967–2977. [CrossRef]
40. Xiang, M.; Cheung, S.C.P.; Tu, J.Y.; Zhang, W.H. A multi-fluid modelling approach for the air entrainment and internal bubbly flow region in hydraulic jumps. *Ocean Eng.* **2014**, *91*, 51–63. [CrossRef]
41. Parsi, M.; Agrawal, M.; Srinivasan, V.; Vieira, R.E.; Torres, C.F.; McLaury, B.S.; Shirazi, S.A.; Schleicher, E.; Hampel, U. Assessment of a hybrid CFD model for simulation of complex vertical upward gas–liquid churn flow. *Chem. Eng. Res. Des.* **2016**, *105*, 71–84. [CrossRef]
42. Zahedi, P.; Zhang, J.; Arabnejad, H.; McLaury, B.S.; Shirazi, S.A. CFD simulation of multiphase flows and erosion predictions under annular flow and low liquid loading conditions. *Wear* **2017**, *376–377*, 1260–1270. [CrossRef]
43. Akhlaghi, M.; Mohammadi, V.; Nouri, N.M.; Taherkhani, M.; Karimi, M. Multi-Fluid VOF model assessment to simulate the horizontal air–water intermittent flow. *Chem. Eng. Res. Des.* **2019**, *152*, 48–59. [CrossRef]
44. Denèfle, R.; Mimouni, S.; Caltagirone, J.-P.; Vincent, S. Multifield hybrid approach for two-phase flow modeling—Part 1: Adiabatic flows. *Comput. Fluids* **2015**, *113*, 106–111. [CrossRef]
45. Chen, G.; Wang, Q.; He, S. Assessment of an Eulerian multi-fluid VOF model for simulation of multiphase flow in an industrial Ruhrstahl–Heraeus degasser. *Metall. Res. Technol.* **2019**, *116*, 617. [CrossRef]
46. Shonibare, O.Y.; Wardle, K.E. Numerical investigation of vertical plunging jet using a hybrid multifluid–VOF multiphase CFD Solver. *Int. J. Chem. Eng.* **2015**, *2015*, 1–14. [CrossRef]
47. Cerne, G.; Petelin, S.; Tiselj, I. Coupling of the interface tracking and the two-fluid models for the simulation of incompressible two-phase flow. *J. Comput. Phys.* **2001**, *171*, 776–804. [CrossRef]
48. Meng, L.; Gao, S.; Wei, D.; Zhao, Q.; Cui, B.; Shen, Y.; Song, Z. Particulate flow modelling in a spiral separator by using the Eulerian multi-fluid VOF approach. *Int. J. Min. Sci. Technol.* **2023**, *33*, 251–263. [CrossRef]
49. Hou, D.; Zhao, Q.; Cui, B.; Wei, D.; Song, Z.; Feng, Y. The effects of solid shear stress and wall roughness on the simulation of cylindrical hydrocyclone classification. *Powder Technol.* **2021**, *393*, 184–204. [CrossRef]
50. Gidaspow, D. *Multiphase Flow and Fluidization: Continuum and Kinetic Theory Descriptions*; Harcourt Brace & Company Academic Press: San Diego, CA, USA, 1994; ISBN 978-0-12-282470-9.
51. Hou, D.; Cui, B.; Zhao, Q.; Wei, D.; Song, Z.; Feng, Y. Research on the structure of the cylindrical hydrocyclone spigot to mitigate the misplacement of particles. *Powder Technol.* **2021**, *387*, 61–71. [CrossRef]
52. Huo, J.; Wang, Z.; Luan, X.; Jing, M.; Hou, S.; Jiang, J.; Zhang, B. The CFD modeling of bund overtopping phenomena and prediction of dynamic pressure on the bund. *J. Loss Prev. Process. Ind.* **2022**, *74*, 104653. [CrossRef]
53. Pattanapol, W.; Wakes, S.J.; Hilton, M.J.; Dickinson, K.J.M. Modeling of surface roughness for flow over a complex vegetated surface. In Proceedings of the World Academy of Science, Engineering and Technology, Canakkale, Türkiye, 14–16 December 2007; Volume 26, pp. 273–281.
54. Cebeci, T.; Bradshaw, P. *Momentum Transfer in Boundary Layers*; Hemisphere Publishing Corporation: Washington, DC, USA, 1977; ISBN 978-0-07-010300-3.
55. Schlichting, H.; Gersten, K. *Boundary-Layer Theory*, 9th ed.; Springer: Berlin, Germany, 2017; ISBN 978-3-662-52919-5.
56. Adams, T.; Grant, C.; Watson, H. A simple algorithm to relate measured surface roughness to equivalent sand-grain roughness. *Int. J. Mech. Eng. Mechatron.* **2012**, *1*, 66–71. [CrossRef]
57. Gao, S.; Meng, L.; Wei, D.; Zhao, Q.; Cui, B.; Zhou, X. A Method for Predicting and Optimising Mineral Particle Separation Indexes in Spiral Concentrator Flow Film. CN Patent CN115270587A, 9 July 2023.
58. Mishra, B.K.; Tripathy, A. A preliminary study of particle separation in spiral concentrators using DEM. *Int. J. Miner. Process.* **2010**, *94*, 192–195. [CrossRef]
59. Lee, S.; Stokes, Y.; Bertozzi, A.L. Behavior of a particle-laden flow in a spiral channel. *Phys. Fluids* **2014**, *26*, 043302. [CrossRef]

minerals

MDPI

Article

Optimization of the Matrix in a Transverse-Field High-Gradient Magnetic Separator for an Improved Ilmenite Separation

Jianguo Liu [1,2,3,4], Huixin Dai [1,*], Lili Yu [2,3,4], Chenghang Wang [2,3,4,5], Jiaying Feng [2,3,4], Peilun Li [2,3,4,5] and Shaohua Xu [2,3,4]

1 Faculty of Land Resources Engineering, Kunming University of Science and Technology, Kunming 650093, China; liujianguo@grre.gd.cn
2 Institute of Resources Utilization and Rare Earth Development, Guangdong Academy of Sciences, Guangzhou 510650, China
3 State Key Laboratory of Separation and Comprehensive Utilization of Rare Metals, Guangzhou 510650, China
4 Guangdong Key Laboratory of Mineral Resources Development and Comprehensive Utilization, Guangzhou 510650, China
5 Southwest United Graduate School, Kunming 650092, China
* Correspondence: dhx6688@sina.com

Abstract: Transverse-field high-gradient magnetic separators (HGMSs) are an important complement to longitudinal-field HGMSs in mineral processing due to their several advantages. However, the processing capacity of the transverse-field HGMS is smaller than that of the longitudinal-field HGMS. Consequently, research on the optimization of the matrix box for improving the processing capacity is essential. This work investigates the optimization of the matrix box for the SSS® HGMS to enhance the ilmenite separation efficiency and processing capacity. The results show that the matrix's influence on separation performance is primarily influenced by the diameter of the rod matrix, the filling ratio, the depth of the matrix in the direction of slurry flow, and the ore unloading efficiency. Ilmenite pre-concentration tests are carried out using a test sample ore from Panzhihua, China. Pilot-scale validation research is carried out. The test results indicate that the depth of the matrix box should not be considerably thick, as an excessive number of layers increases the capture zone, but simultaneously reduces the unloading efficiency. The depth of the matrix box should neither be considerably thick nor particularly thin, as this would result in low processing capacity. Meanwhile, the segmented multi-layer matrix box should be used to balance the capturing and unloading performance. Finally, an optimal double-layer matrix ring is applied to the industrial transverse-field HGMS, and its inner and outer rings are equipped with matrix boxes with ϕ3 mm and ϕ2 mm rods, respectively, which improves its pre-concentrate efficiency and processing capacity. The concentrate indexes of the transverse-field HGMS is achieved with a TiO_2 grade of 18.01% and a recovery of 87.28%, which is better than the separation indexes of the longitudinal-field HGMS.

Keywords: transverse magnetic field; high-gradient magnetic separator; magnetic matrix; ilmenite; separation efficiency

1. Introduction

High-gradient magnetic separation is one of the most effective physical separation techniques, and high-gradient magnetic separators (HGMSs) have been widely used for separating essential metal ores, removing magnetic impurities of non-metallic ores, and eliminating suspended solids and other impurities in wastewater or waste gas [1–6]. The longitudinal-field HGMS [7] and transverse-field HGMS [8] are typical HGMSs in industrial applications. The transverse-field HGMS serves as an important complement to the longitudinal-field HGMS in mineral processing due to its several advantages, such as reduced susceptibility to corrosion of the magnetic system, lower probability of magnetic matrix blockages, and longer service life. To date, the main transverse-field HGMSs include the SSS-I series wet high-intensity magnetic separator (WHIMS), the SSS-II series double-frequency double-vertical-rings pulsating WHIMS [9], the SSS-III series double-frequency double-layer vertical-rings pulsating WHIMS [10], and the SSS-IV series horizontal-ring non-pulsating WHIMS. The SSS-II series transverse-field HGMS has been used for limonite processing and ilmenite processing in China. The SSS-II-1750 transverse-field HGMS has been used for limonite processing at Nanjing Meishan Metallurgical Development Co., Ltd., Najing, China. Mining Brach since 2009; this system has successfully resolved the issues of magnetic system corrosion and magnetic matrix blockage of the original longitudinal-field HGMS [11]. The SSS-II-2000 transverse-field HGMS has been used for ilmenite processing at Pangang Group Mining Co., Ltd., Panzhihua, China. Titanium Dressing Plant since 2012. The concentrate grade is 1.58%, and the concentrate recovery is 5.08% more than the indexes achieved using the longitudinal-field HGMS [9].

The magnetic field in a transverse-field HGMS is oriented horizontally. In contrast with the longitudinal-field HGMS through industrial application practice, the slurry and magnetic systems in a transverse-field HGMS are independently configured, preventing direct contact between the slurry and the magnetic system. Consequently, the magnetic system is protected from corrosion, leading to a longer service life [12]. Furthermore, the magnetic matrix in a transverse-field HGMS is arranged in a rectangular lattice, offering a smoother slurry flow path compared to the rhombic lattice arrangement in the longitudinal-field HGMS. This configuration reduces the probability of particle sedimentation on the magnetic matrix, thereby preventing blockages in the matrix box. Consequently, the capturing and unloading performance are improved, and the durability of the matrix box is enhanced [13].

However, the width of the sorting ring and the volume of the matrix box are limited by the magnetic system's spacing, resulting in a reduced processing capacity. The equipment must be improved to increase the processing capacity. Numerous factors related to the magnetic matrix still influence the separation performance of transverse-field high-gradient magnetic separation, including the matrix's shape, the diameter of the rod matrix, filling ratio, the depth of the matrix in the direction of slurry flow, and the discharge efficiency of the captured particles [14–20]. Professors Luzheng Chen and Dongfang Lu have conducted extensive and fruitful work on HGMS. This work presents the findings of a study on matrix box optimization in transverse-field HGMSs to improve the processing capacity for ilmenite pre-concentration. Titanium is an important element in ilmenite, which has been extensively used in the fields of aerospace, petrochemical, and automobile manufacturing [21]. Ilmenite in China is considered low grade, with grades typically ranging from 5% to 10%, necessitating pre-concentration. Magnetic separation is a key pre-concentration method, and HGMS has been widely used in the beneficiation of ilmenite. Lu et al. [22] studied a novel method for the efficient recovery of ilmenite by HGMS coupling with magnetic fluid. Chen et al. [23] used a new type HGMS (CHGMS) for ilmenite separation. In their study, ilmenite from Panzhihua was used to conduct pilot-scale validation research and an industrial application for the pre-concentration of ilmenite. The matrix box was improved, the processing capacity was significantly increased, and the ore unloading rate and separation index were increased.

2. Materials, Equipment, and Methods

2.1. Experimental Sample

The experimental sample consists of tailings from the separation of vanadium–titanium magnetite (Panzhihua, Sichuan, China) as the primary ilmenite. The TiO_2 content of the sample is relatively low, and its mineral composition is complex. The results of the multi-elemental analysis and particle size distribution for this sample are presented in Tables 1 and 2, respectively. The data in the tables indicate that the TiO_2 grade of the sample is 9.90%, which is relatively low, thereby necessitating a pre-concentration process. The predominant particle size of the sample is +0.074 mm, with a corresponding yield of 70.57%. This result shows that the sample has a moderate particle size distribution, neither excessively coarse nor fine. Additionally, the liberation degree of the sample is 90.70%, which exceeds 90%, illustrating an excellent level of mineral liberation suitable for the pre-concentration process.

Table 1. Multi-elemental analysis of the experimental sample (%).

Element	TiO_2	TFe	V_2O_5	SiO_2	Al_2O_3	CaO	MgO	S
Content	9.90	28.08	0.19	29.60	9.74	5.27	12.17	0.12

Table 2. Particle size distribution of the experimental sample (%).

Grain Size (mm)	Yield	TiO_2 Grade	TiO_2 Proportion	Liberation Degree
+0.25	6.04	8.49	5.18	44.39
−0.25 + 0.20	7.86	8.81	6.99	87.07
−0.20 + 0.15	18.50	10.15	18.97	88.27
−0.15 + 0.10	20.16	10.11	20.59	92.18
−0.10 + 0.074	18.01	9.92	18.05	95.63
−0.074 + 0.043	11.55	12.35	14.41	97.95
−0.043 + 0.038	11.23	8.91	10.11	99.08
−0.038	6.65	8.51	5.70	99.25
Total	100.00	9.90	100.00	90.70

The mineral composition analysis, the equilibrium distribution of TiO_2 in the minerals, and the dissemination relationship of ilmenite in some large mineral particles are illustrated in Figure 1, Figure 2, and Figure 3, respectively. The data presented in these figures reveal that ilmenite, a valuable mineral, constitutes 13.58% of the experimental sample, with 70.42% of the TiO_2 distributed within it. The theoretical TiO_2 grade of ilmenite is 51.35%, which is slightly lower than the value of 52.66% derived from the chemical formula. This discrepancy may be attributed to the lattice substitution effects from magnesium, manganese, and other impurity elements. The primary gangue minerals identified in the sample include plagioclase, olivine, titanaugite, chlorite, and quartz, with contents of 24.62%, 15.17%, 14.05%, 10.99%, and 10.68%, respectively. The undissociated ilmenite is mainly intergrown with amphibole and chlorite or Ti-magnetite, which would interfere with the TiO_2 grade of pre-concentrate of HGMS. Ilmenite belongs to the group of weakly magnetic minerals with relatively strong magnetic properties, and most of them can be captured using a high-gradient magnetic matrix box and enter the magnetic products under the background magnetic field strength of 0.4~0.6 T.

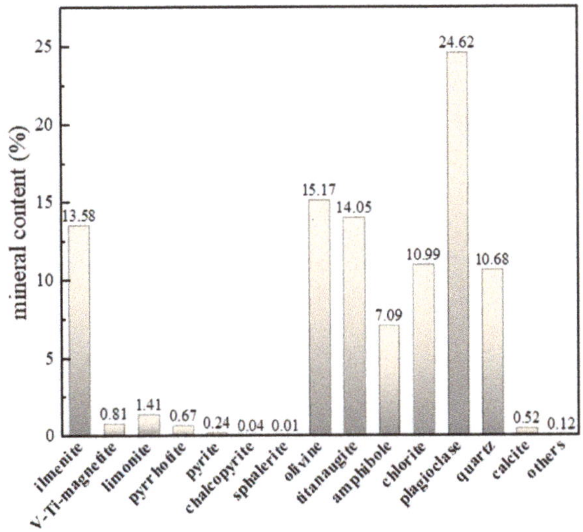

Figure 1. Mineral composition analysis of the experimental sample.

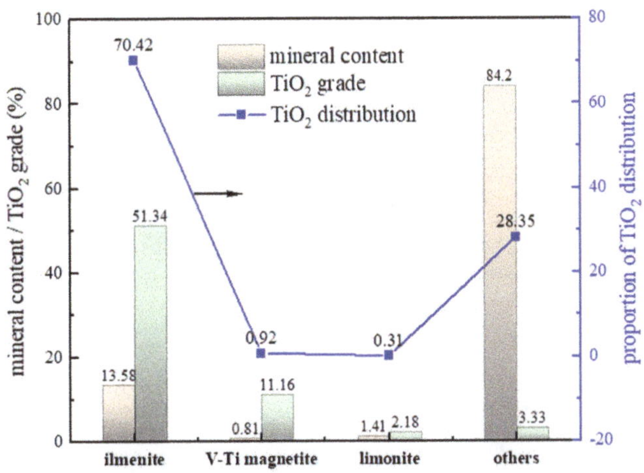

Figure 2. Equilibrium distribution of TiO_2 in these minerals.

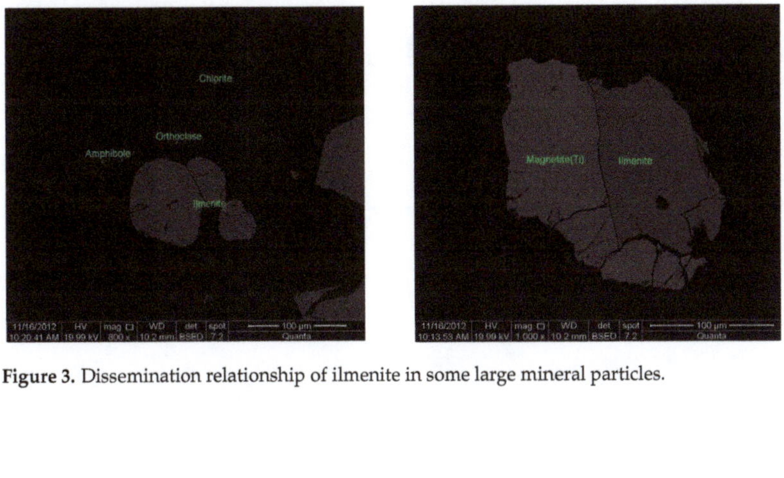

Figure 3. Dissemination relationship of ilmenite in some large mineral particles.

2.2. Material Properties of the Magnetic Matrix

The conventional matrix used for separating metal ores in HGMSs is a rod with a circle cross-section, and its diameter is typically 3, 2, or 1.5 mm. In this work, the magnetic matrix material from Dalian F.T.Z. Special Steel Product Factory (Dalian, China) is selected for the laboratory investigation and industrial application research. The steel type is Cr10NiTiMo, and its chemical composition and mechanical properties are shown in Table 3; the B–H testing curve of the magnetic matrix Cr10NiTiMo is shown in Figure 4.

Table 3. Chemical composition contents and tensile strength of the magnetic matrix.

Element	Fe	C	Si	Mn	P	S
Content (%)	88.86	0.010	0.260	0.080	0.016	0.004
Element	**Cr**	**Ni**	**Mo**	**Ti**	**Tensile Strength (MPa)**	
Content (%)	10.100	0.240	0.210	0.220	880,980	

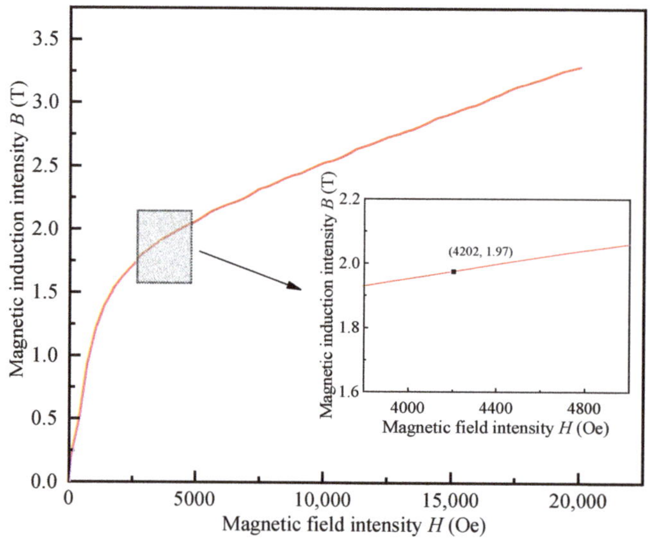

Figure 4. B–H testing curve of magnetic matrix Cr10NiTiMo.

2.3. Pilot-Scale and Industrial-Scale Transverse-Field HGMS

The pilot-scale separator used is the SSS-180 × 80 × 250 cyclic transverse-field HGMS, with an effective separation chamber volume of 180 × 80 × 250 mm^3. The structural diagram and equipment photograph are provided in Figure 5. The entire separation principle of this device is similar to that of the SLon cyclic pulsating HGMS, with the primary distinction being the orientation of the magnetic system.

The industrial-scale separator used is a typical SSS vertical ring transverse-field HGMS. The structure of this separator is shown in Figure 6. The equipment is primarily composed of a pulsating mechanism, an iron yoke, vertical rings, a gas–water combined ore unloading device, a magnetic matrix box, a ring driver, exciting coils, and magnetic poles.

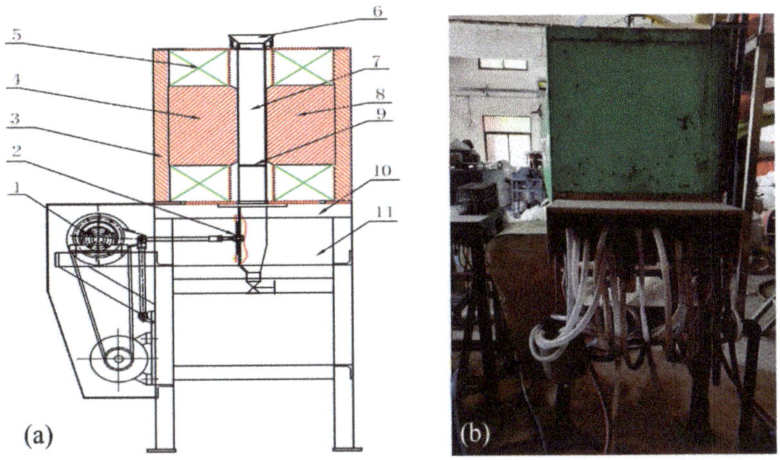

Figure 5. Equipment structure diagram (**a**) and physical diagram (**b**). 1—Gearing; 2—pulsating tympanic membrane; 3—magnetic system; 4—left magnetic pole; 5—exciting coils; 6—feed bin; 7—separation chamber; 8—right magnetic pole; 9—magnetic matrix box holder; 10—equipment frame body; 11—cooling water inlet and outlet area.

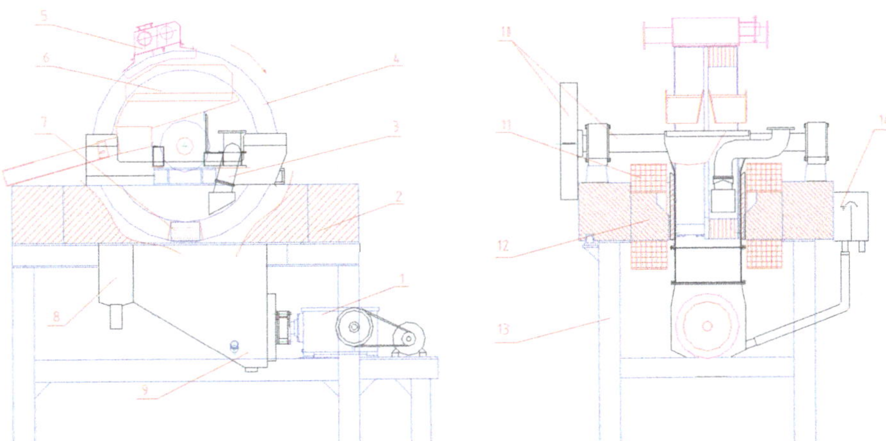

Figure 6. Structure of a typical SSS vertical ring transverse-field HGMS. 1—Pulsating mechanism; 2—iron yoke; 3—feeding pipe; 4—rotating rings; 5—gas–water combined unloading device; 6—concentrate bucket; 7—magnetic matrix box; 8—middling hopper; 9—tailings hopper; 10—ring driver; 11—excitation coils; 12—magnetic pole; 13—rack; 14—liquid level bucket.

The slurry, introduced through the feed pipe, passes an arc-shaped separation zone formed by opposing magnetic poles. Within this zone, magnetic mineral particles are attracted to and captured on the surface of the matrix. Meanwhile, non-magnetic particles, unaffected by the magnetic field, pass through the matrix units and enter into the tailings hopper. The pulsating mechanism, with adjustable frequency and stroke, creates an up-and-down scouring motion in the slurry to avoid matrix blockage, which aids in loosing and removing gangue that has deposited on the matrix surface, facilitating its flow into the tailings hopper. The remaining magnetic particles, still adsorbed on the matrix, rotate with the matrix ring and gradually exit the magnetic field. In the top region without magnetic field, a gas–water combined unloading device is utilized to wash the magnetic particles off the matrix surface, which are then collected in the concentrate hopper, forming the magnetic product. This process can effectively separate magnetic particles from non-magnetic particles.

The design of the transverse-field HGMS effectively isolates the magnet from the ore slurry, thereby preventing abrasion and corrosion caused by the high-speed pulp. This isolation simplifies the maintenance of the matrix rings, allowing for the detached ring to be easily hoisted for servicing. A key feature of this separator is the "moving-cleaning" capability of the magnetic matrix, which significantly reduces the risk of incrustation on the matrix's surface and delays matrix blockage. In practical applications, the feed always contains strongly magnetic particles, such as magnetite and tramp iron. These particles alternately move between the capture and discharge steps, as illustrated in Figure 7. When the matrix ring rotates into the arc-shaped magnetic zone, magnetic particles adsorb to both sides of the rod matrix. The weakly magnetic particles are flushed into the concentrate bucket as the ring moves to the unloading area. Meanwhile, the strongly magnetic particles remain trapped at the bottom of the rod matrix due to residual magnetism. Upon returning to the magnetic zone, the strongly magnetic particles are re-mobilized to a horizontal position, preventing the formation of a fixed deposition. This continuous movement of the strong magnetic particles effectively cleans the rod matrix's surface, preventing magnetic minerals from staying in one place and without moving, which can be described as a "moving-cleaning" process. Thus, no deposition occurs on the surface of the rod matrix, and the pulp path remains unblocked.

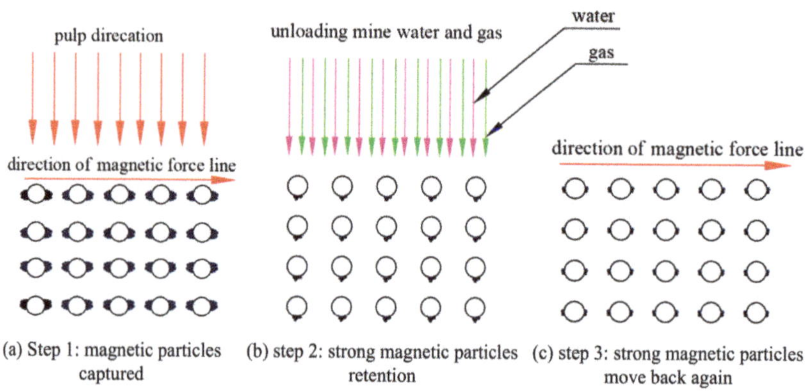

Figure 7. Process of magnetic matrix capture–washing–recapture (moving–cleaning).

2.4. Methods of Matrix Box Optimization

As previously mentioned, the factors that affect the separation performance and processing capacity of the magnetic matrix mainly include the diameter of the rod matrix, the filling ratio, and the depth of the matrix in the direction of slurry flow. The rod diameter is closely related to the distribution of the magnetic field within the background magnetic field. The matrix filling ratio affects the capture capacity and the efficiency of ore unloading. The depth of the matrix in the direction of slurry flow affects the smoothness of slurry flow, thereby influencing the processing capacity and concentrate indexes.

The relationship between the rod diameter and the mineral particle size is typically expressed as Equation (1), according to the research result [19] and test experience:

$$K = \frac{D_a}{D_b} > 10 \quad (1)$$

where K is a specific value, D_a is the rod diameter, and D_b is the particle size of the recoverable minerals.

The ore unloading efficiency is defined by the following equation:

$$U = \frac{Q_0 - Q_1}{Q_0} \times 100\% \quad (2)$$

where U is the ore unloading efficiency, Q_0 is the total quantity of concentrate adsorbed by the matrix, and Q_1 is the residual concentrate quantity remaining in the matrix after ore unloading.

The recovery of particle-sized fraction in the concentrate is a critical factor in assessing the separation performance of the magnetic matrix. This factor is defined as follows:

$$\xi = \frac{\varepsilon \times \gamma}{\kappa} \times 100\% \quad (3)$$

where ξ is the recovery of the particle-sized fraction, ε is the recovery of concentrate, γ is the metal proportion of a certain size of fraction in the concentrate, and κ is the metal proportion of a certain size of fraction in feeding.

3. Experimental Results and Discussion

3.1. Magnetic Matrix Optimization

The particle size fraction of the sample is primally distributed in −0.25 + 0.10 mm and −0.10 + 0.043 mm, and the corresponding yields are 46.52% and 40.79%, respectively. Consequently, the suitable rod diameters are more than 2.5 and 1.0 mm because $K > 10$. The ϕ3.0 mm rod is suitable as the matrix for separating the −0.25 + 0.10 mm fraction. Meanwhile, the optimal rod diameter for separating the −0.10 + 0.043 mm fraction remains to be determined. The maximum magnetic induction strength measured on the surface of rods with different diameters under varying exciting currents is presented in Figure 8. The test data are measured in the unmoving state of the device.

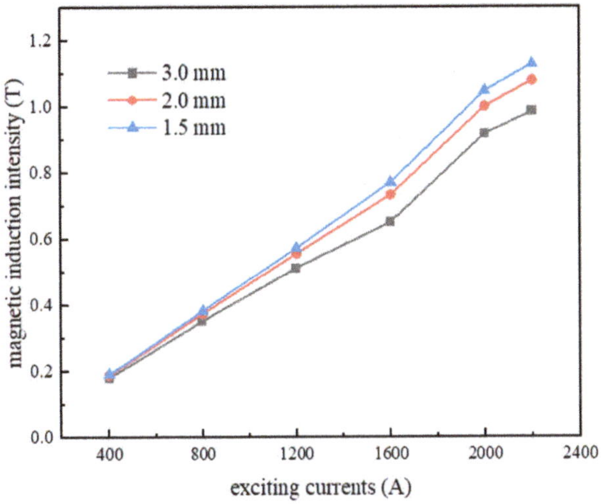

Figure 8. Maximum magnetic induction strength measured on the surface of rods with different diameters under varying exciting currents.

The results presented in Figure 8 indicate that the difference in magnetic induction intensity on the surfaces of rods with different diameters becomes pronounced with the increase in the excitation current. Specifically, the magnetic induction field of the ϕ3.0 mm rod is lower than that of the ϕ2.0 and ϕ1.5 mm rods. Under a large current, the magnetic induction intensities of the ϕ2.0 and ϕ1.5 mm rods are similar, with the former slightly lower. Considering the stability of the mechanical properties and the induction field, the ϕ3.0 mm magnetic matrix rod is suitable for separating coarse particle size fractions with stronger magnetism, and ϕ2.0 and ϕ1.5 mm are suitable for separating relatively fine particle size fractions with weaker magnetism. This inference can be verified by the models of particle tracking with different diameters of rod matrixes in Figure 9, as well as the capture probability of particles using rod matrixes in Figure 10.

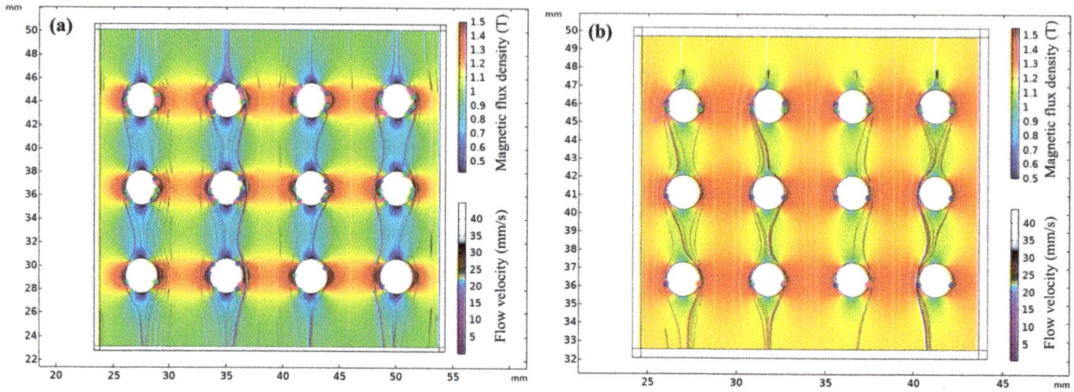

Figure 9. Particle tracking models for rod matrixes with diameters of ϕ3 mm (**a**) and ϕ1.5 mm (**b**). Note: the background magnetic field intensity is 0.42 T, the particle size is −0.25 + 0.10 mm and −0.10 + 0.043 mm, the relative permeability of the particle is 1.006, and the initial velocity is 40 mm/s.

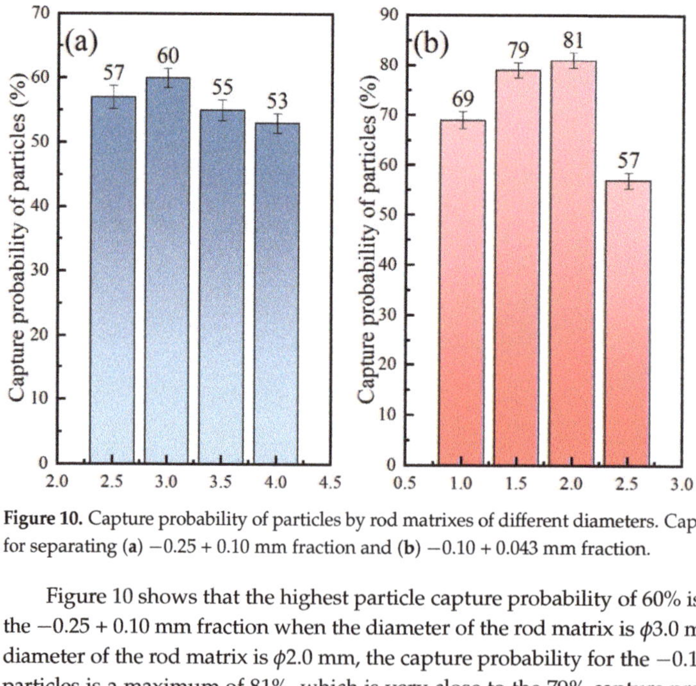

Figure 10. Capture probability of particles by rod matrixes of different diameters. Capture probability for separating (**a**) −0.25 + 0.10 mm fraction and (**b**) −0.10 + 0.043 mm fraction.

Figure 10 shows that the highest particle capture probability of 60% is achieved for the −0.25 + 0.10 mm fraction when the diameter of the rod matrix is ϕ3.0 mm. When the diameter of the rod matrix is ϕ2.0 mm, the capture probability for the −0.10 + 0.043 mm particles is a maximum of 81%, which is very close to the 79% capture probability for a diameter of ϕ1.5 mm.

In addition to the diameter of the rods, their filling ratio also affects the magnetic field distribution. Taking the magnetic matrix rod of ϕ2.0 mm as an example, the maximum magnetic induction strength measured on the surfaces of rods with different filling ratios under varying exciting currents is shown in Figure 11.

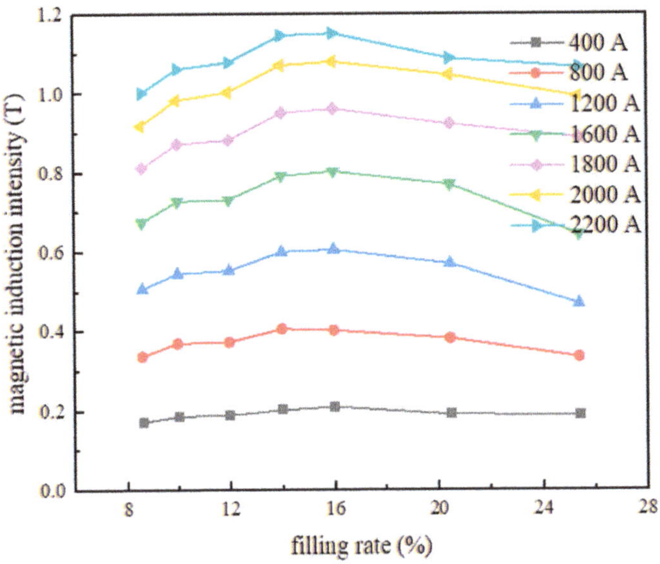

Figure 11. Maximum magnetic induction strength measured on the surfaces of rods with different filling ratios under varying exciting currents.

Figure 11 demonstrates that the magnetic induction strength initially rises with the increase in the filling ratio and then declines. The filling ratio always falls in a range from 14% to 16% when the maximum magnetic induction intensity is achieved at various excitation currents. A higher magnetic induction intensity corresponds to a stronger ability to capture magnetic minerals. Therefore, an optimal filling ratio of approximately 16% is chosen.

The number of arranged layers (also referred to as "depth") in the magnetic matrix is closely correlated with the capture capacity and the unloading efficiency of magnetic particles. A typical matrix box utilized in the transverse-field HGMS is depicted in Figure 12, with a depth of 248 mm, a length of 140 mm, a width of 78 mm, and a rod diameter of 2.0 mm as an example. The depth of the matrix box is high, resulting in a large processing capacity. The optimal test conditions are as follows: feed quantity of 1500 g, stroke of 14 mm, pulsating frequency of 240 r/min, and magnetic field intensity of 0.42 T, which are obtained through condition tests. The experimental results of the ore unloading efficiency with different flushing water quantities are presented in Table 4. The findings show that ore unloading efficiency ranges from 75.59% to 85.78% as the flushing water flow rate increases from 3 L/min to 12 L/min. Despite the higher flushing water volumes, the ore unloading efficiency remains suboptimal. This notion suggests that the commonly used magnetic matrix box, characterized by a considerable depth, is inherently limited in its ability to achieve high ore unloading efficiency.

Figure 12. Typical matrix box in a pilot-scale transverse-field HGMS with a diameter of 2 mm.

Table 4. Experimental results of the ore unloading efficiency with different flushing water quantities.

Flushing Water (L/min)	Production	Yield (%)	TiO_2 Grade (%)	TiO_2 Recovery (%)	Q_0 (g)	Q_1 (g)	U (%)
3	Concentrate	41.26	16.85	70.23	618.90	151.10	75.59
	Tailings	58.74	5.02	29.77			
	Total	100.00	9.90	100.00			
6	Concentrate	43.65	17.16	75.66	654.75	123.41	81.15
	Tailings	56.35	4.28	24.34			
	Total	100.00	9.90	100.00			
9	Concentrate	44.26	17.22	76.99	663.90	101.56	84.70
	Tailings	55.74	4.09	23.01			
	Total	100.00	9.90	100.00			
12	Concentrate	44.85	17.31	78.42	672.75	95.67	85.78
	Tailings	55.15	3.87	21.58			
	Total	100.00	9.90	100.00			

The element matrix method is utilized to assess the capture of magnetic particles at various positions within the matrix box (from top to bottom) and further investigate the potential for reducing the depth of the matrix box. In this method, the matrix box is divided into five elements (A~E) (Figure 13). The steps to obtain the amount of ilmenite particles captured by each layer of matrix (A~E) are as follows: restore the process of capturing ilmenite particles in the magnetic matrix box in the stationary state; when the feeding is finished, keep the original background magnetic field strength unchanged (0.42 T) until there is no liquid outflow at the bottom of the matrix box; and then reduce the background magnetic field strength to 0.1 T, although, at this time, the ilmenite particles are subjected to the magnetic force of the substrate decreases, but the particles are still captured on the surface of the substrate under the action of the capillary force of the liquid. Immediately after that, perform a layer-by-layer slow removal of the matrix and collection of ilmenite particles in the order of A~E. Multiple experiments were conducted, and the data were statistically processed. The experimental results, summarized in Table 5, suggest a gradual decrease in the capture weight and the TiO_2 grade as one moves from element A to E. Specifically, the capture weight and TiO_2 grade of element A are 252.60 g and 16.19%, whereas those of element E are only 55.65 g and 11.82%, respectively. These findings indicate that the capture efficiency in the lower elements of the matrix box is significantly reduced, thus supporting the feasibility of minimizing the number of layers in the matrix box.

Figure 13. Magnetic matrix box divided into five elements (**A**~**E**).

Table 5. Experimental results using the element matrix method.

Matrix Box Element	Production	Weight (g)	Yield (g)	TiO_2 Grade (%)	TiO_2 Recovery (%)
A	Concentrate1	252.60	16.84	16.19	27.55
B	Concentrate2	258.90	17.26	18.24	31.81
C	Concentrate3	110.25	7.35	16.61	12.34
D	Concentrate4	94.35	6.29	13.93	8.85
E	Concentrate5	55.65	3.71	11.82	4.43
	Tailings	728.25	48.55	3.06	15.01
	Total	1500.00	100.00	9.90	100.00

In summary, Tables 4 and 5 indicate that the matrix box should not be considerably deep, as this would hinder the efficient unloading of ore and prevent the achievement of optimal capture performance. Therefore, the typically used deep, integrated matrix box must be segmented to balance capturing and unloading efficiencies, allowing for independent unloading in each segment.

3.2. Pilot-Scale Validation Research

Segmented short magnetic matrix boxes with rod diameters of 3, 2, and 1.5 mm are fabricated to validate the results of the magnetic matrix box optimization, as illustrated in Figure 14a–c. The typical combinations of these short matrix boxes used in pilot-scale HGMSs are demonstrated in Figure 15. The rod diameter of the upper-layer matrix box is intentionally larger than that of the lower-layer matrix box, facilitating the early capture of relatively coarse and strongly magnetic particles. The upper and lower-layer matrix boxes are connected via a holder (Figure 14d). The experimental results for the typical combination of the double-layer matrix box are presented in Table 6. The combination of matrix boxes with rod diameters of $\phi 3$ and $\phi 2$ mm yields the optimal performance for ilmenite separation. This configuration achieves a TiO_2 grade of 18.46% and a recovery of 87.32%. Additionally, this configuration exhibits the highest ore unloading efficiency, at 94.07%. The separation index and ore unloading efficiency surpass the results presented in Table 4 that were obtained using the traditional integrated matrix box. The suitable magnetic matrix box combination is ϕ3 mm and ϕ2 mm. This notion means that the two-layer ring can be used in industrial equipment. The inner ring is configured with a magnetic matrix box of ϕ3 mm rod. Meanwhile, the outer ring is configured with a magnetic matrix box of ϕ2 mm rod.

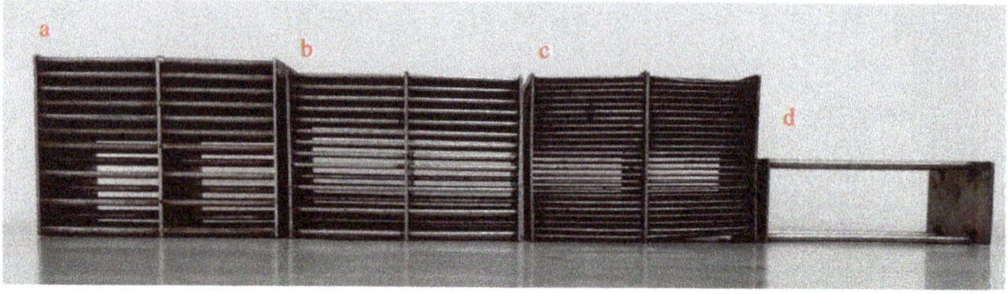

Figure 14. Magnetic matrix boxes and a holder: (**a**)—ϕ3 mm rod; (**b**)—ϕ2 mm rod; (**c**)—ϕ1.5 mm rod; (**d**)—holder.

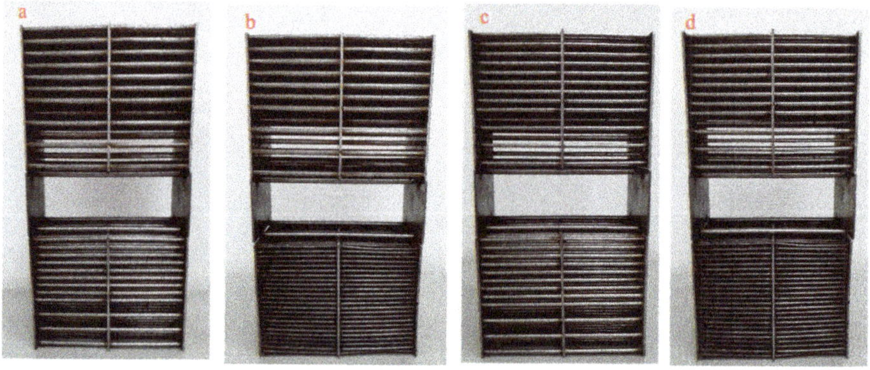

Figure 15. Segmented magnetic matrix boxes with different rod diameter combinations: (**a**)—ϕ3 mm and ϕ2 mm; (**b**)—ϕ3 mm and ϕ1.5 mm; (**c**)—ϕ2 mm and ϕ2 mm; (**d**)—ϕ2 mm and ϕ1.5 mm.

Table 6. Experimental results of the typical combination of the double-layer matrix boxes.

Combination Form of the Box	Production	Yield (%)	TiO_2 Grade (%)	TiO_2 Recovery (%)	Q_0 (g)	Q_1 (g)	U (%)
ϕ3 and ϕ2 mm	Concentrate	46.82	18.46	87.32	702.30	41.63	94.07
	Tailings	53.18	2.36	12.68			
	Total	100.00	9.90	100.00			
ϕ3 and ϕ1.5 mm	Concentrate	44.83	17.35	78.55	672.45	61.57	90.84
	Tailings	55.17	3.85	21.45			
	Total	100.00	9.90	100.00			
ϕ2 and ϕ2 mm	Concentrate	45.30	18.31	83.76	679.50	51.38	92.44
	Tailings	54.70	2.94	16.24			
	Total	100.00	9.90	100.00			
ϕ2 and ϕ1.5 mm	Concentrate	43.88	17.26	76.48	658.20	68.25	89.63
	Tailings	56.12	4.15	23.52			
	Total	100.00	9.90	100.00			

The specific results of the screen analysis for the optimal concentrate and size fraction recovery are listed in Table 7. These results indicate that the recovery of the size fractions ranging from −0.15 mm to +0.038 mm is relatively uniform, with a value of approximately 90%. However, the recovery of the coarser fraction (+0.15 mm) and the finer fraction (−0.038 mm) is comparatively low. The lower recovery of the coarser fraction can be attributed to poor dissociation. Meanwhile, the reduced recovery of the finer fraction is due to the insufficient magnetic attraction resulting from the small particle size. Therefore, these pilot-scale experimental findings provide strong evidence that supports the validity of the magnetic matrix optimization.

Table 7. Particle size distribution in the concentrate and size fraction recovery.

Size Fraction (mm)	Yield (%)	TiO_2 Grade (%)	TiO_2 Proportion (%)	Size Fraction Recovery (%)
+0.25	6.26	13.17	4.47	75.27
−0.25 + 0.20	8.18	13.56	6.01	75.05
−0.20 + 0.15	18.43	17.22	17.19	79.12
−0.15 + 0.10	20.75	19.51	21.93	92.99
−0.10 + 0.074	17.11	20.64	19.13	92.53
−0.074 + 0.043	12.19	23.19	15.31	92.78
−0.043 + 0.038	10.86	17.56	10.33	89.21
−0.038	6.22	16.76	5.65	86.49
Total	100.00	18.46	100.00	87.32

3.3. Industrial Application Case

The industrial testing machine is developed based on the success of the pilot test. The two-layer sorting ring for the transverse-field industrial HGMS is demonstrated in Figure 16. The volumes of the magnetic matrix boxes in the outer and inner rings presented in Table 8. The rod diameter of the outer ring is 2 mm, while that of the inner ring is 3 mm. The data in Table 8 indicate that the magnetic matrix volumes of the transverse-field and longitudinal-field HGMSs are comparable when using the same vertical ring diameter. The transverse-field industrial HGMS with the two-layer ring used in the beneficiation plant is shown in Figure 17.

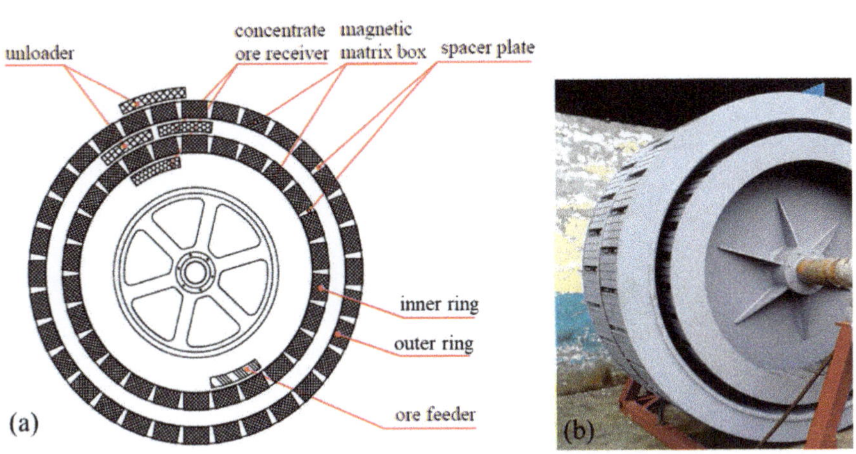

Figure 16. Diagram of the two-layer ring (**a**) and physical image (**b**).

Table 8. Volume of the magnetic matrix boxes of the HGMSs.

Type of HGMS	Model	Magnetic Rings	Box Size (L × W × H) (mm)	Box Number	Volume (m^3)	Total (m^3)
Transverse magnetic field HGMS	2750	Outer	330 × 174 × 226	80	1.04	1.65
		Inner	330 × 156 × 184	64	0.61	
Longitudinal magnetic field HGMS	2750	One	300 × 182 × 174	176	1.67	1.67

Figure 17. Industrial equipment of the SSS-III-2750 two-layer ring transverse-field HGMS used in the beneficiation plant.

The industrial best operating parameters of the transverse-field HGMS (SSS-III-2750) are as follows: vertical ring speed of 4.0 r/min, magnetic field intensity of 0.45 T, pulsating stroke of 10 mm, and pulsating frequency of 120 r/min. The industrial best operating parameters of the longitudinal-field HGMS are as follows: vertical ring speed of 4.0 r/min, magnetic field intensity of 0.50 T, pulsating stroke of 14 mm, and pulsating frequency of 170 r/min. The comprehensive index for a 1-month stable production is provided in Table 9. The result indicates that the separation index for the transverse-field HGMS outperforms that of the longitudinal-field HGMS. The production achieved a TiO_2 grade of 18.01% with a recovery of 87.28% after a roughing separation. The particle size distribution in the concentrate and size fraction recovery of the two-layer ring transverse-field HGMS is shown in Table 10. The size fraction recovery is more than 91% when the size fraction is from −0.15 mm to +0.043 mm, similar to the result in the pilot plant test.

Table 9. Industrial test results of the two-layer ring transverse- and longitudinal-field HGMSs.

Type of HGMS	Model	Production	Yield	TiO_2 Grade	TiO_2 Recovery
Transverse magnetic field HGMS	2750 (two-layer ring)	Concentrate	47.98	18.01	87.28
		Tailings	52.02	2.42	12.72
		Total	100.00	9.90	100.00
Longitudinal magnetic field HGMS	2750	Concentrate	50.61	16.73	85.49
		Tailings	49.39	2.91	14.51
		Total	100.00	9.90	100.00

Table 10. Particle size distribution in the concentrate and size fraction recovery of the two-layer ring transverse-field HGMS.

Size Fraction (mm)	Yield (%)	TiO_2 Grade (%)	TiO_2 Proportion (%)	Size Fraction Recovery (%)
+0.25	6.52	12.77	4.62	77.95
−0.25 + 0.20	8.76	13.12	6.38	79.74
−0.20 + 0.15	18.93	16.85	17.72	81.54
−0.15 + 0.10	20.52	19.07	21.73	92.17
−0.10 + 0.074	16.94	20.16	18.97	91.76
−0.074 + 0.043	11.89	22.89	15.12	91.59
−0.043 + 0.038	10.13	17.48	9.83	84.94
−0.038	6.31	16.06	5.63	86.22
Total	100.00	18.01	100.00	87.28

4. Conclusions

The transverse-field vertical ring HGMS is a suitable industrial magnetic separator used in the ilmenite separation of pre-concentrations. This separator can be used as an important supplement to the longitudinal-field vertical ring HGMS in mineral processing due to its features, including no magnet wear and corrosion and the natural magnetic matrix "moving-cleaning". However, the processing capacity should be improved. This work carried out the optimization of a matrix box in pilot-scale and industrial transverse-field HGMSs to improve the pre-concentrate efficiency and processing capacity of ilmenite. The obtained conclusions are as follows.

The diameter of the rod matrix, the filling ratio of the rod matrix, and the depth of the matrix box influence the pre-concentrate's efficiency and processing capacity. The processing capacity can be increased with the increase in the depth of the matrix box. The depth of a single matrix box should not be considerably deep, as this would result in a low ore unloading rate, which can be divided into different thinner matrix boxes, for example, two-layer matrix boxes with ϕ3 and ϕ2 mm rods combined, which can improve the pre-concentrate's efficiency and processing capacity.

A two-layer sorting ring is fabricated in the industrial transverse-field HGMS based on the research results. The inner ring is configured with a matrix box of a ϕ3 mm rod, which is used to separate the coarse fraction products. Meanwhile, the outer ring is configured with a matrix box of ϕ2 mm rod, which is used to separate the fine fraction products. The industrial test results indicate that the pre-concentrate efficiency and processing capacity are improved. Moreover, the separation index of the transverse-field HGMS is better than that of the longitudinal-field HGMS.

Author Contributions: Investigation, Data curation, Formal analysis, Writing—original draft, J.L.; Conceptualization, Supervision, Methodology, Writing—review & editing, H.D.; Analysis, Methodology, L.Y.; Supervision, Writing—review & editing, C.W.; Methodology, Validation, J.F.; Methodology, Resources, P.L.; Software. S.X. All authors have read and agreed to the published version of the manuscript.

Funding: This research work was financially supported by the Yunnan Provincial Science and Technology Project at Southwest United Graduate School (Grant No. 202302AO370002), National Key Research and Development Program (Grant No. 2023YFC2908101), Natural Science Foundation of Guangdong Province (Grant No. 2022A1515012165), Special Fund Project of Guangdong Academy of Sciences (No. 2022GDASZH-2022010104).

Data Availability Statement: Data available on request from the authors.

Conflicts of Interest: The authors declare no conflicts of interest.

References

1. Jian, Z.; Xiong, T.; Fan, Y.; Lu, C. Selective Capture of Magnetic Wires to Particles in High Gradient Magnetic Separation. *Minera* **2019**, *9*, 509. [CrossRef]
2. Zhi, H.; Jian, L.; Li, H.; Yu, W.; Dong, L.; Xia, Z.; Zi, X. Dymanic particle accumulation on a single wire in transverse field pulsatir high gradient magnetic separator. *Miner. Eng.* **2022**, *183*, 107609.
3. Wen, L.; Yue, H.; Rui, X.; En, G. A Preliminary Investigation into Separating Performance and Magnetic Field Characterist Analysis Based on a Novel Matrix. *Minerals* **2018**, *8*, 94. [CrossRef]
4. Chen, L.; Liao, G.; Qian, Z.; Chen, J. Vibrating high gradient magnetic separation for purification of iron impurities under dı condition. *Int. J. Miner. Process.* **2012**, *130*, 136–140. [CrossRef]
5. Karapinar, N. Magnetic separation of ferrihydrite from wastewater by magnetic seeding and high-gradient magnetic separatio *Int. J. Miner. Process.* **2003**, *71*, 45–54. [CrossRef]
6. Hu, Z.; Liu, J.; Gan, T.; Lu, D.; Wang, Y.; Zheng, X. High-intensity magnetic separation for recovery of $LiFePO_4$ and graphite fro spent lithium-ion batteries. *Sep. Purif. Technol.* **2022**, *297*, 121486. [CrossRef]
7. Zeng, W.; Xiong, D. The latest application of SLon vertical ring and pulsating high-gradient magnetic separator. *Miner. Eng.* **200** *16*, 563–565. [CrossRef]
8. Yu, T. Development of SSS-II wet type double frequency pulsation double vertical ring high gradient magnetic separator. *Mε Mine* **2004**, *3*, 37–39. (In Chinese) [CrossRef]
9. Feng, W.; Shi, T.; Ming, Z.; Qing, D. Development of SSS-II horizontal magnetic field high gradient magnetic separator and i application in Panzhihua Midi titanium plant. *Met. Mine* **2014**, *2*, 122–126. (In Chinese) [CrossRef]
10. Shao, X.; Yu, T.; Jun, Z.; Tian, L. Development and application of SSS-III series vertical ring pulsating HGMS with horizonta magnetic field. *Met. Mine* **2019**, *5*, 137–141. (In Chinese)
11. Yu, T.; Jun, Z.; Wernham, J. The research on application of new technologies in high gradient magnetic separation with horizonta magnetic line. In Proceedings of the XXV International Mineral Processing Congress (IMPC) 2010 proceedings, Brisbane, Australi 6–10 September 2010; pp. 1281–1286.
12. Yu, W.; Zi, X.; Xia, Z.; Dong, L.; Zi, S. Matching relation between matrix aspect ratio and applied magnetic induction for maximur particle capture in transversal high gradient magnetic separation. *Miner. Eng.* **2020**, *151*, 106316.
13. Zhi, H.; Dong, L.; Xia, Z.; Yu, W.; Zi, X.; Shao, X. Development of a high-gradient magnetic separator for enhancing selectiv separation: A review. *Powder Technol.* **2023**, *421*, 118435.
14. Zhi, H.; Jun, Z.; Jian, L.; Yu, T.; Xia, Z. Model of particle accumulation on matrices in transverse field pulsating high gradien magnetic separator. *Miner. Eng.* **2020**, *146*, 106105.
15. Xia, Z.; Yu, W.; Dong, L.; Xu, L.; Si, L.; Hao, C. Comparative study on the performance of circular and elliptic cross-sectio matrices in axial high gradient magnetic separation: Role of the applied magnetic induction. *Miner. Eng.* **2017**, *110*, 12–19.
16. Xia, Z.; Yu, W.; Dong, L.; Xu, L. Theoretical and experimental study on elliptic matrices in the transversal high gradient magneti separation. *Miner. Eng.* **2017**, *111*, 68–78.
17. Chen, L.; Liu, W.; Zeng, J.; Ren, P. Quantitative investigation on magnetic capture of single wires in pulsating HGMS. *Powde Technol.* **2017**, *313*, 54–59. [CrossRef]
18. Zeng, J.; Tong, X.; Ren, P.; Chen, L. Theoretical description on size matching for magnetic element to independent particle in hig gradient magnetic separation. *Miner. Eng.* **2019**, *135*, 74–82. [CrossRef]
19. Zheng, X.; Wang, Y.; Lu, D. Study on capture radius and efficiency of fine weakly magnetic minerals in high gradient magneti field. *Miner. Eng.* **2015**, *74*, 79–85. [CrossRef]
20. Xia, Z.; Shuang, L.; Li, D.; Zi, S.; Hai, T.; Yu, W.; Dong, L. Investigation of particle capture by grooved plates with elliptic teeth fo high intensity magnetic separation. *Sep. Purif. Technol.* **2023**, *325*, 124685.
21. Xia, Z.; Li, D.; Shuang, L.; Zi, J.; Dong, L.; Ke, J.; Kadir, C.; Bing, P.; Yu, W. A novel method for efficient recovery of ilmenite by high gradient magnetic separation coupling with magnetic fluid. *Miner. Eng.* **2023**, *202*, 108279.
22. Xia, Z.; Xing, L.; Li, D.; Sheng, Z.; Yu, W.; Dong, L. Clean and short flowsheet of recovering ultrafine ilmenite by high gradien magnetic separation coupling with magnetic fluid. *Miner. Eng.* **2025**, *222*, 109156.
23. Jian, Z.; Lu, C.; Ruo, Y.; Xiong, T.; Peng, R.; Yong, Z. Centrifugal high gradient magnetic separation of fine ilmenite. *Int. J. Mine Process.* **2017**, *168*, 48–54.

minerals

MDPI

ticle

ffect of the Cross-Sectional Geometry of the Mixed article Zone on the Spiral Separation Process and Its tructural Optimization

uling Gao *, Qian Wang *, Xiaohong Zhou, Chunyu Liu, Yanbai Shen and Baoyu Cui

School of Resources and Civil Engineering, Northeastern University, Shenyang 110819, China
* Correspondence: gaoshuling@mail.neu.edu.cn (S.G.); wangqian@stumail.neu.edu.cn (Q.W.); Tel.: +86-13709818446 (S.G.); +86-15940412486 (Q.W.)

Abstract: Cross-sectional geometry is of significance in determining the flow characteristics and the particles separation in spiral concentrators. The effects of the cross-sectional geometry within the mixed particle zone on the secondary flow and the separation process of hematite and quartz particles with a size of 89.5 μm were investigated via a computational fluid dynamics (CFD) approach. The optimization of the line segment slopes was conducted using response surface methodology (RSM). The results indicate that the peak radial fluxes and average radial velocities of the secondary flow are positively correlated with the corresponding line segment slope. The independent adjustment of line slopes in regions I and II, and the interaction of line slopes in regions I and III, influence the separation efficiency of hematite and quartz particles significantly. The separation performance of the experimental spiral concentrator with a cross-sectional profile of the optimized line segments for a feed of hematite and quartz with a size range of −100 + 75 μm is remarkably improved by nearly 5%. This study provides insights for the cross-section design of spiral concentrators for the effective separation of coarse-grained hematite and quartz.

Keywords: cross-sectional geometry; mixed particle zone; separation efficiency; radial flux; response surface methodology (RSM); difference in migration amount of particles

1. Introduction

The construction of green mines represents an important initiative to promote the development of an environmentally friendly approach within the mining sector and to enhance the efficiency of comprehensive resource utilization. The spiral concentrator is an example of thin-flowing film mineral processing equipment. The operation of the spiral concentrator is based on the combined action of gravity and the centrifugal inertia force. Given its array of superior characteristics, the spiral concentrator has been extensively employed in the separation of a diverse range of materials, including iron, tungsten, tin, tantalum, niobium, coal, seashore sand, and other nonferrous metals, rare metals, and nonmetallic minerals with density differences [1–6].

The superior separation performance of the spiral concentrator is contingent upon the differences in the properties of valuable mineral and gangue mineral particles, as well as the unique flow field characteristics [7,8]. The slurry flow in the spiral concentrator rotates in a downwards direction along the trough under the influence of gravity, forming a longitudinal mainstream. Concurrently, it moves in a radial direction due to the centrifugal inertia force. Following a period of development and evolution, the flow ultimately exhibits the characteristics of a coexistence of mainstream and secondary flow, as well as a coexistence of laminar, transitional, and turbulent flow [9]. The structural parameters of the spiral concentrator exert the most important effect on the evolution characteristics of the flow field, which in turn has a profound effect on the migration and distribution of the particles attached to the internal flow field. Once the parameters have been determined, it is challenging to adjust them during the production process [10–13]. Because optimization of the production index by adjusting the operating parameters can obtain only a slight improvement, careful selection of the structural parameters of the spiral concentrator is vital. In recent years, researchers have investigated the effects of the structural parameters, such as the pitch, cross-sectional profile, and splitter position of spiral concentrators, on the separation efficiency in order to increase the universality of spiral concentrators in the rapid separation of mineral particles with different properties in a short process [14–18]. It has been demonstrated that a gentler cross-section shape is more appropriate for the separation of fine-grained minerals, vice versa Pitch exerts a comprehensive influence on the predominant velocity of the slurry, the thickness of the film, and the movement of the particles. A reduction in pitch can facilitate the establishment of a more stable separation flow field, whereas an increase has a more pronounced impact on high-density particles, which is conducive to accelerating the formation of particle bands [10]. Nevertheless, advances in the elucidation of the mechanism of the effects of the structural parameters on hydrodynamics and their influence on the separation process for particles with different properties from a theoretical standpoint have been relatively limited.

Cross-sectional geometry is a crucial structural parameter that influences several key hydrodynamic parameters, such as the flow film thickness, velocity of the primary flow, secondary flow, and Reynolds number. Commonly employed cross-sectional curves are derived from cubic parabolas, elliptic curves, or combinations of curvilinear functions [19,20]. It can be postulated that any material to be separated will have a corresponding optimal cross-sectional geometry. However, generating large quantities of cross-sectional curves by modifying the functional expression and then selecting the optimal cross-sectional curves suitable for different materials is nearly impossible. The process of screening through trial-and-error tests incurs a significant cost in labor, material, and financial resources. Although the cross-sectional curves used by the spiral concentrators in current production or research have the potential to achieve satisfactory separation efficiency, the separation results have not been interpreted in terms of particle migration processes. Additionally, the design and optimization of the localized geometric parameters of the mixed particle zone are rarely considered. In light of the aforementioned considerations, it is urgently necessary to perform a rational optimization of the structural parameters of the spiral concentrator, with a particular focus on the cross-sectional geometrical parameters.

The current methods employed in the research on the optimization of cross-sectional geometry for spiral concentrators can be broadly classified into three main categories: theoretical studies, physical inspection and measurements, and numerical simulations. Theoretical research has, to varying degrees, overlooked the interactions between the particles and has only considered simplified particle motion, resulting in significant differences between the results predicted by the theoretical models and the experimental results. Laser doppler anemometry (LDA) [21], particle image velocimetry (PIV), and positron emission particle tracking (PEPT) [22–24], among other flow and particle monitoring devices, enable the acquisition of fluid and particle motion information without interference in the flow field. Nevertheless, these methods show some drawbacks, such as the use of cumbersome detection processes, strict test conditions, and high maintenance costs. Numerical simulations employing such methods as the discrete element method (DEM), smoothed particle hydrodynamics (SPH), and computational fluid dynamics (CFD) have been successfully employed to simulate the flow field and particle motion of spiral concentrators [25,26]. Among these, SPH is a meshless simulation approach with high accuracy [20], which requires significant computational volume and shows low computational efficiency in fluid simulations. The DEM accurately determines the force, spatial position, velocity, and energy changes of particles [27]. However, it only considers particle motion and fails to account for the interaction between the fluid and the particles, as well as between the particles in the spiral concentrator slurry system. The bidirectional coupling of the DEM and CFD leads to a significant increase in the computational cost. Due to the limitations of the DEM and SPH methods, CFD is a more frequently employed approach in simulations of spiral concentrators.

The flow field within the spiral concentrator is characterized as an open channel flow. Therefore, it is essential to select an appropriate multiphase flow model for accurate simulation of the gas–liquid–solid three-phase system in a spiral concentrator. Currently, the simulation methods applied to the particle flow in spiral concentrators are based primarily on the Eulerian–Lagrange method and the Eulerian–Eulerian method. The Eulerian–Lagrange method treats the fluid and the particles as the continuous phase and the discrete phase, respectively. The discrete phase model (DPM) is commonly used to monitor the movement of the particles in a spiral concentrator. To minimize the computational cost, interactions between the particles and the impact of the particles on the continuous phase are typically disregarded in the DPM model. Consequently, this approach is limited to predicting multiphase flows with volume fractions below 12%, restricting its applicability to high-concentration feed conditions. By contrast, the Euler–Euler method, which views particles as a proposed fluid within a continuous medium and characterizes particle interactions through the kinetic theory of granular flow (KTGF) model, provides an effective and viable approach for predicting highly concentrated particle flows in a spiral concentrator from the perspective of fluid dynamics [28,29].

This study introduces the concept of a localized optimization design for the cross-sectional geometry of the mixed particle zone of a spiral concentrator. A numerical analysis was conducted using a three-turn spiral concentrator geometry model to optimize the structure. The effects of the cross-sectional geometry on the secondary flow characteristics and particle separation processes were investigated. Additionally, the mechanism of the effect of the line segment slopes within the mixed particle zone on the separation performance of the spiral concentrator was elucidated. Moreover, the response surface analysis method was employed to explore the interaction between different line segment slopes and to identify the optimal parameter configuration for the localized cross-sectional geometry. The effectiveness of the localized adjustment strategy for line segment slopes was confirmed through actual separation texts focusing on specific materials.

2. Experimental Equipment and Methods

2.1. *Structural Parameters of the Spiral Concentrator and Adjustment Strategy for the Localized Cross-Sectional Geometry*

2.1.1. Structure Parameters of the Spiral Concentrator

A laboratory-scale spiral concentrator with a diameter of 300 mm was employed as the research object for our numerical simulation research. The relevant structural parameters are presented in Table 1.

Table 1. Relevant structural parameters of the spiral concentrator.

Structural Parameters	Value
Outer radius (R)/mm	150
Inner radius (r_0)/mm	30
Pitch (P)/mm	240
Number of turns (N)	3.25

The expression for the sectional curve function of the spiral concentrator is as follows:

$$|x| = m|y|^n \tag{1}$$

where m is the coefficient of the sectional curve function and n is the exponent of the sectional curve function.

The function image in the third quadrant of the Cartesian coordinate system was selected for the cross-sectional geometry of the spiral concentrator, as illustrated in Figure 1. The trough slope β is defined as the acute angle formed by the tangent line at a radial position r of the cross-sectional curve and the horizontal line. The downwards bevel angle, designated γ, is the acute angle formed by the connecting line between the endpoints of the inner and outer edges of the spiral concentrator and the horizontal line. It is calculated using the following formula:

$$\tan\gamma = \left[m^{1/n}(R - r_0)^{1-1/n}\right]^{-1} \tag{2}$$

where R is the outer radius and r_0 is the inner radius. In this study, the functional expression for the cross-section curve of the laboratory-scale spiral concentrator has m = 0.01748, n = 3, and "γ" = 9°.

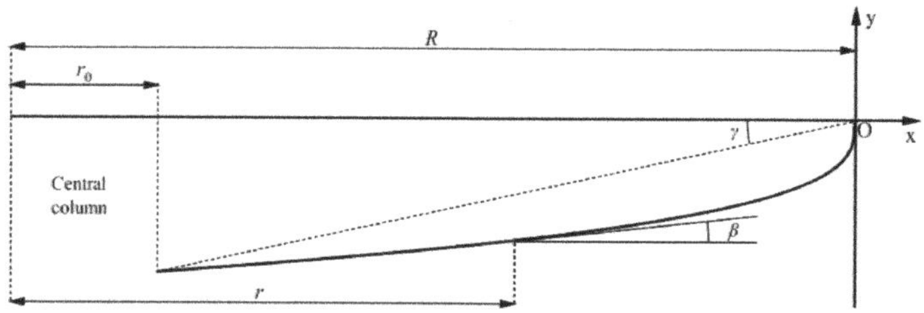

Figure 1. Schematic of the cross-sectional geometry of the spiral concentrator.

2.1.2. Configuration of the Sectional Curve Function and Downwards Bevel Angle

Initially, geometric models of the spiral concentrator were constructed according to downwards bevel angles of 3°, 6°, 9°, 12°, and 15°, along with the cross-sectional curve function (refer to Table 2), to simulate the flow field and particle motion behaviour, respectively. Single-factor and factorial experiments were then carried out by numerical simulation to analyze the individual and combined effects of the downwards bevel angle and cross-sectional curve function on the separation efficiency of quartz and hematite. The findings are illustrated in Figure 2.

Table 2. Functional expression of the sectional curve of the spiral concentrator.

Numbers of Sectional Curve Function	Functional Expression for Sectional Curve
Equation (1)	$|x| = 1.44824|y|^{1.5}$
Equation (2)	$|x| = 0.33220|y|^{2}$
Equation (3)	$|x| = 0.07620|y|^{2.5}$
Equation (4)	$|x| = 0.01748|y|^{3}$
Equation (5)	$|x| = 0.00401|y|^{3.5}$

The single-factor test results, depicted in Figure 2a,b, show that adjusting the downwards bevel angle and the exponent of the cross-sectional curve function within a specific range significantly affects the separation efficiency of hematite and quartz particles. When the downwards bevel angle is set at 6° or 9° and the cross-sectional curve function has an index of 2.5–3.5, the peak separation efficiency can reach approximately 80%, and the separation efficiency is enhanced. The factorial experiment results in Figure 2c reveal that the optimal separation performance occurs at a 6° downwards bevel angle and a cross-sectional curve function index of 2.5, achieving a maximum separation efficiency of 82.02%. This configuration is identified as the optimal setup for the cross-sectional curve function and the downwards bevel angle. The control equation for the cross-sectional curve at this point is given by the following Equation (3):

$$|x| = 0.212|y|^{2.5} \tag{3}$$

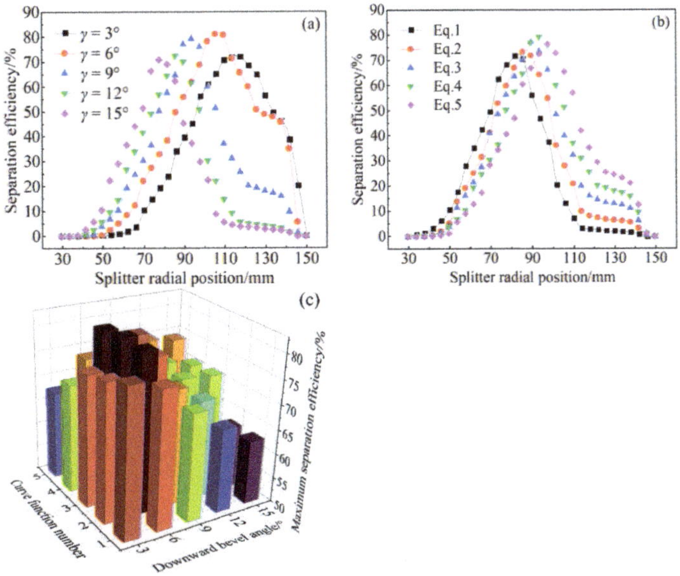

Figure 2. Effects of the cross-sectional geometry parameters on the separation efficiency. (**a**) Effect of the downwards bevel angle; (**b**) effect of the sectional curve function; (**c**) interaction between the downwards bevel angle and the sectional curve function.

2.1.3. Delimiting the Mixed Particle Zone in the Trough and a Strategy for Localized Adjustment of Line Segment Slopes

The above-described configuration allows the study of the distribution of the yield and separation efficiency of hematite and quartz in the trough at the end of the third turn. As illustrated in Figure 3, hematite and quartz are selectively distributed and concentrated in the trough, with the maximum separation efficiency occurring at r = 100 mm. Notably, a higher yield of mixed quartz is found near r = 70 mm, whereas a greater proportion of hematite is mixed in the quartz accumulation region at r = 100–115 mm. The separation efficiency curve peaks sharply, with a significant decrease in the radial region of r = 100–130 mm, leading to a narrow adaptation range for the splitter. For these reasons, the radial region of r = 70–130 mm is identified as the mixed particle zone.

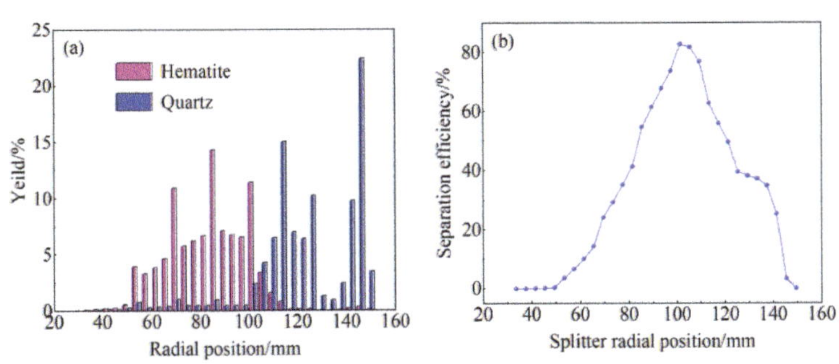

Figure 3. Yield distribution and separation efficiency of hematite and quartz at the end of the third turn. (**a**) Yield Distribution; (**b**) separation efficiency.

As illustrated in Figure 4, the curve of the mixed particle zone is divided into three regions: region I (r = 70–90 mm), region II (r = 90–110 mm), and region III (r = 110–130 mm). The characteristic radial positions for these regions are 80, 100, and 120 mm, respectively. The slopes of the curves in these regions are relatively small, allowing the line segments between the endpoints to approximate the original curves, resulting in the creation of three line segments: P_1P_2, P_2P_3, and P_3P_4.

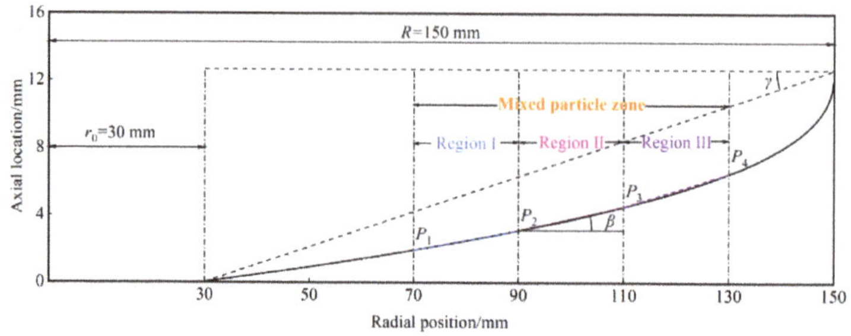

Figure 4. Resegmentation of the cross-sectional geometry of the mixed particle zone.

The geometrical models of the spiral concentrator were constructed based on specified line segment slopes provided in Table 3. Then, the effects of the adjustments of these slopes on the secondary flow characteristics and particle separation processes were investigated. Response surface methodology (RSM) is a systematic approach for solving multivariate problems through experimental design, including central composite design (CCD) and Box–Behnken design (BBD). BBD is efficient for predicting the main effects, interactions, and squared terms with fewer trials than CCD [30]. Consequently, in this paper, a BBD experiment with three factors and three levels was conducted to establish a relationship between the variables and the responses. Optimization of the line segment slopes in the mixed particle zone was then carried out.

Table 3. Adjustment values for the line segment slopes in the localized region.

Radial Region	Line Segments	Radial Range	Slop of Curve/°	Line Segment Slope/°
Region I	P_1P_2	r = 70–90 mm	3.34	1; 3; 5; 7; 9; 11
Region II	P_2P_3	r = 90–110 mm	4.09	1; 2; 3; 4; 5; 6; 7
Region III	P_3P_4	r = 110–130 mm	5.62	0; 2; 4; 6; 8; 10

2.2. Feeding Properties

The experimental material consisted of hematite and quartz in Anshan-type hematite ores, with densities of 4950 kg/m^3 and 2650 kg/m^3, respectively. To simplify the numerical calculations for simulations of full-size particles, both minerals were set at a median particle size (D_{50}) of 89.5 µm, reflecting the particle composition of the feed for the spiral concentrator at the Qidashan iron ore processing plant, as illustrated in Figure 5. The feed consists of hematite and quartz particles at a mass ratio of 1.87, corresponding to an iron grade of 45.59%. The feed has a solid mass concentration of 20 wt% and a volumetric flow rate of 0.48 m^3/h (8 L/min). It should be noted that the particle size range of hematite and quartz used for the practical separation and structural optimization of spiral concentrators in this study is −100 + 75 µm. The suitability of this research methodology and structural optimization method for separating mineral particles with a wide size range or other density differences still requires extensive validation.

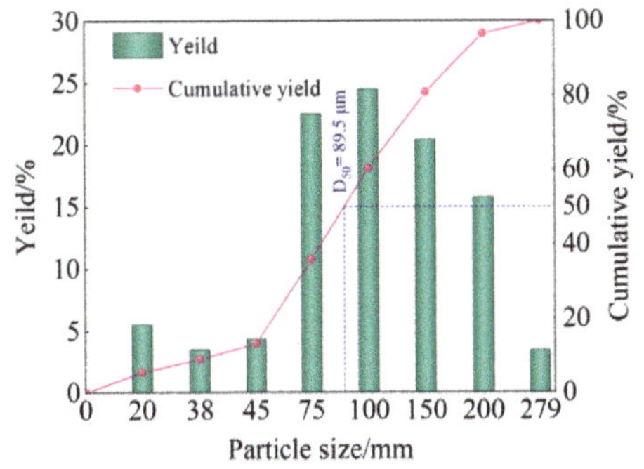

Figure 5. Particle size composition of the feed to the spiral separator at the Qidashan Iron Mine Processing Plant.

2.3. Numerical Model and Simulation Conditions

The RNG k-ε turbulence model demonstrates favorable predictive accuracy for vortex flows with relatively large curvature and is suitable for predicting low Reynolds number flows in the near-wall region. Consequently, the RNG k-ε turbulence model is adopted to describe the turbulence characteristics of the fluid in the spiral concentrator.

The Eulerian multifluid VOF model, which incorporates the Bagnold effect, is employed to simulate the intricate slurry system of gas–liquid stratified flows and solid–liquid mixed flows in a spiral concentrator. This model provides a framework to couple the VOF and Euler–Euler multiphase models to simulate two-phase interfaces in stratified flows and predict particle motion in mixed flows [7,19,31]. The VOF interface tracking model is used to describe the gas–liquid stratified flow pattern, and the continuity and momentum equations for the incompressible two-phase flow can be expressed in the following Equation (4) and Equation (5), respectively:

$$\frac{\partial \rho u_i}{\partial x_i} = 0 \tag{4}$$

$$\frac{\partial}{\partial x_j}(\rho u_i u_j) = -\frac{\partial p}{\partial x_i} + \rho g_i + \frac{\partial}{\partial x_j}\left[\mu_{eff}\left(\frac{\partial u_i}{\partial x_j} + \frac{\partial u_j}{\partial x_i}\right)\right] \tag{5}$$

where ρ represents the fluid density, u_{ij} is the mean fluid velocity, P is the static pressure, μ_{eff} is the effective viscosity (molecular μ plus turbulent u_t), and g is the acceleration due to gravity.

For the Euler–Euler model, the continuous and discrete phases are regarded as interpenetrating continua. Particles with different densities or sizes were treated as separate phases. The phase-averaged continuity and momentum equations of each phase are given by the following Equations (6) and (7):

$$\frac{\partial}{\partial t}(\alpha_q \rho_q) + \nabla \cdot \left(\alpha_q \rho_q \vec{v}_q\right) = 0 \tag{6}$$

$$\frac{\partial}{\partial t}\left(\alpha_q \rho_q \vec{v_q}\right) + \nabla \cdot \left(\alpha_q \rho_q \vec{v}_q \vec{v}_q\right) = -\alpha_q \nabla p - \nabla p_q + \nabla \cdot \overline{\overline{\tau}}_q + \alpha_q \rho_q \vec{g} + \vec{F}_d + \vec{F}_B \tag{7}$$

where q represents different phases (q = 1 for water, 2 for air, and $3 - n$ for the different types of particles); α_q is the volume fraction of phase q; $\vec{v}_q$ is the velocity vector of phase q; ρ_q is the material density of phase q; $\vec{g}$ is the gravitational acceleration; p_q ($q \geq 3$) is the qth solid pressure described by the model of Syamlal–O'Brien [32]; p is the pressure shared by all phases; $\vec{F}_d$ is the drag force [33]; $\vec{F}_B$ is the Bagnold lift force [34]; $\bar{\bar{\tau}}_q$ is the qth phase stress–strain tensor and is expressed as follows:

$$\bar{\bar{\tau}}_q = \alpha_q \mu_q \left(\nabla \vec{v}_q + \nabla \vec{v}_q^{\,T} \right) + \alpha_q \left(\delta_q - \frac{2}{3} \mu_q \right) \nabla \cdot \vec{v}_q \bar{\bar{I}} \tag{8}$$

where μ_q represents the shear viscosity of phase q, which is determined by the correlation of Gidaspow et al. [35]; T is the transposition of matrix; $\bar{\bar{I}}$ is the identity matrix; δ_q is the bulk viscosity of phase q, and is described by the correlations of Lun et al. [36]. The virtual mass force and turbulent dispersion force are not considered during the simulation process. The interphase forces are composed primarily of traction and lift. The Gidaspow model and the symmetric model are employed to address liquid–solid traction and gas–liquid traction, respectively. Bagnold empirical formulas are applied in the lift model and integrated into the lift coefficient module through the user-defined function (UDF) interface [31,34].

A mesh independence test is essential for enhancing the accuracy of the numerical results and optimizing computational efficiency. Since the computational domain mesh remains consistent for each turn, the computational domain mesh of the single-turn spiral concentrator is discretized using ANSYS ICEM 18.0, as depicted in Figure 6a [19]. The mesh of the cross-section is presented in Figure 6b, and it can be seen that the mesh near the trough was fined locally. The minimum height of the bottom wall mesh is 0.15 mm, while the lowest y^+ value is 2. The effects of three meshing methods, with mesh quantities of 78,000, 149,760, and 303,072, respectively, on the surface tangential velocity of the flow film are examined, as illustrated in Figure 7 [19]. The findings indicate that when the number of meshes exceeds 149,760, further increasing the mesh count has a negligible effect on the surface tangential velocity. Considering both computational speed and accuracy, the entire 3.25-turn spiral concentrator is meshed into 486,720 high-quality hexahedral cells, following the second meshing method. The spiral concentrators of various cross-sectional geometry investigated in this research are meshed using the optimized meshing methods mentioned above accordingly.

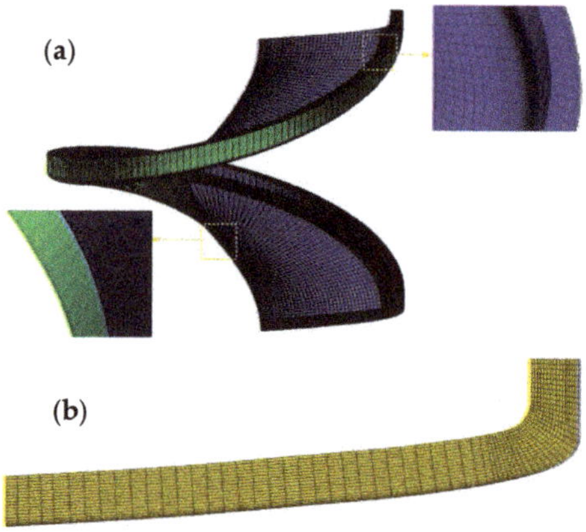

Figure 6. Computational domain mesh of spiral separator. (**a**) Mesh of one turn; (**b**) mesh of the cross-section.

The computational domain of the spiral concentrator is subdivided into four regions: the inlet, upper surface, wall, and outlet. The inlet boundary is defined as a velocity inlet, with a velocity of 0.147 m/s (equivalent to a volume flow rate of 8 L/min) and a turbulence intensity of 5%, classified as medium turbulence intensity. The outlet boundary is designated as a pressure outlet set to standard atmospheric pressure with an air reflux coefficient of 1. The upper surface and wall are designated as free slip and no slip walls, respectively, with standard wall functions applied to solve the equations near the walls [37].

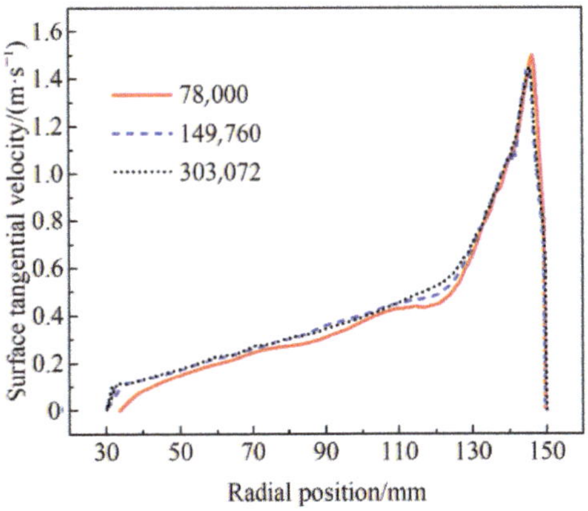

Figure 7. Mesh independence test.

2.4. Data Extraction and Representation

Upon meeting the convergence criterion, secondary flow parameters were extracted for analysis. The secondary flow consists of an inward flow with a negative radial velocity and an outward flow with a positive radial velocity, depending on the flow direction. The inward flow radial flux and the outward flow radial flux are described using the following specific expressions to analyze the development of secondary flow:

$$Q_{in} = S_{in} V_{in,awa} \tag{9}$$

$$Q_{out} = S_{out} V_{out,awa} \tag{10}$$

where S_{in}, S_{out}, $V_{in,awa}$, and $V_{out,awa}$ represent the inward and outward flow region areas and the inward and outward flow fluid area-weighted average radial velocities, respectively. The extraction interval is 1/18 of a turn.

The flow field in a spiral concentrator gradually stabilizes through a specific travelling evolution. The area-weighted average radial velocities of the inward and outward flows are extracted at the end of the third turn within the trough under different conditions to characterize the intensity of the secondary flow.

The particle radial migration behaviour was quantitatively characterized via the cumulative radial flux, which represents the flux of particles in a given radial region from the inlet to a specified trough length. To facilitate comparison, the cumulative radial flux of particles at the midpoint of the three adjusted regions (r = 80, 100, and 120 mm) was extracted at 0.125 turn intervals. Outward migration is indicated by a positive value of the particle radial flux, whereas inward migration is indicated by a negative value.

The mass flow rates of the hematite and quartz particles were extracted under various test conditions. The particle yields in different radial regions were calculated sequentially. Based on these calculations, concentrated products with different distribution widths were obtained by adjusting the position of the splitte. The separation efficiency indicated in Equation (11) [27,38], also called the Schulz's index, may not fully consider the comprehensive quality of the final product, such as in instances of high recovery and low grade. However, due to its intuitiveness, comparability, and quantitative effect, it has sufficient applicability and rationality for determining the separation effect of the two-product system composed of pure minerals of hematite and quartz in this article. Therefore, it is used as a quantitative indicator for the final separation effect of hematite and quartz particles, as follows:

$$E_s = R_v - R_g \tag{11}$$

where E_s is the separation efficiency, R_v is the hematite yield, and R_g is the quartz yield.

3. Results and Discussion

3.1. Effect of Localized Independent Adjustment of the Line Segment Slopes Within the Mixed Particle Zone on the Secondary Flow Characteristics

The flow field of the spiral concentrator transitions from an unsteady to steady state along the trough, with a continuous change in secondary flow. Consequently, it is essential to investigate how localized adjustments to line segment slopes in the mixed particle zone impact secondary flow characteristics.

3.1.1. Effect of Localized Independent Adjustment of the Line Segment Slopes Within the Mixed Particle Zone on the Radial Fluxes of the Secondary Flow

The effects of varying line segment slopes on the radial fluxes of the inward and outward flows in the first turn are shown in Figures 8–10.

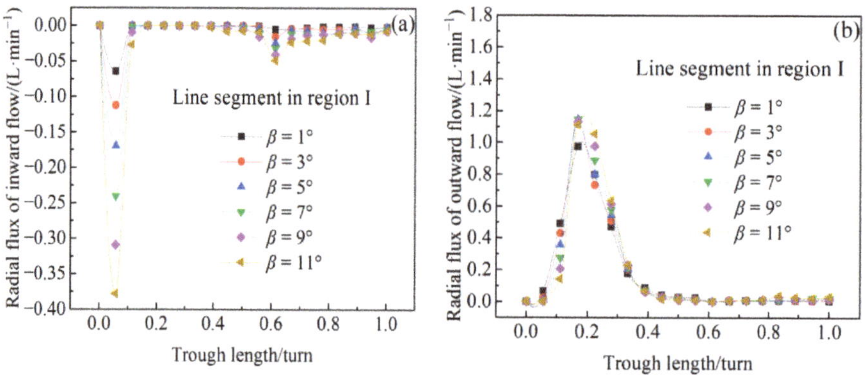

Figure 8. Effect of the line segment slope in region I on the radial fluxes of the inward and outward flows at $r = 80$ mm. (**a**) Radial flux of the inward flow; (**b**) radial flux of the outward flow.

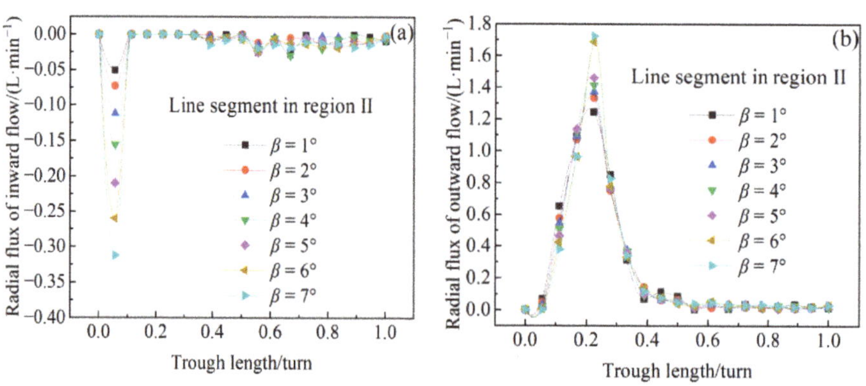

Figure 9. Effects of the line segment slope in region II on the radial fluxes of the inward and outward flows at $r = 100$ mm. (**a**) Radial flux of the inward flow; (**b**) radial flux of the outward flow.

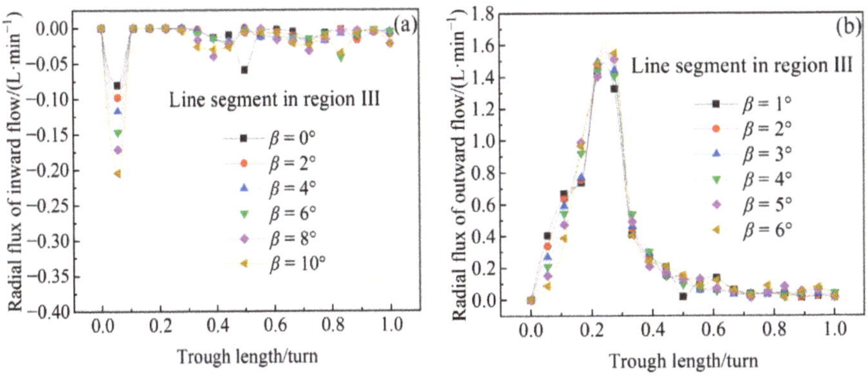

Figure 10. Effects of the line segment slope in region III on the radial fluxes of the inward and outward flows at r = 120 mm. (**a**) Radial flux of the inward flow; (**b**) radial flux of the outward flow.

The results depicted in Figures 8–10 show that adjusting the line segment slopes in the localized region has a relatively consistent effect on the radial flux of the secondary flow in the first turn. The radial flux peaks in the initial stage, then decreases and converges to zero, indicating fluid migration equilibration. The inward flow reaches equilibrium before 0.2 turns, whereas the outward flow requires 0.5 turns. At the initial stage, the peak radial fluxes of the inward and outward flows in distinct regions are positively correlated with the corresponding line segment slopes, with a slightly more pronounced positive correlation observed for the inward flow. These results suggest that increasing the line segment slope can simultaneously increase the flow rates of both the inward and outward flows, thereby facilitating mass exchange within the secondary flow. As the fluid travels longitudinally, the impact of the line segment slope on the radial fluxes of both the inward and outward flows diminishes. Conversely, it is evident that adjusting the line segment slope in region I leads to a more significant peak change in the radial flux of the inward flow, whereas adjusting the slope in region II results in a more pronounced change in the radial flux peaks of the outward flow. Notably, increasing this parameter from 5° to 6° can lead to a peak increase of up to 15.55%. Furthermore, the amplitude of variation in the radial fluxes of the inward and outward flows increases with the outward movement of the radial position.

3.1.2. Effect of Localized Independent Adjustment of the Line Segment Slopes Within the Mixed Particle Zone on the Velocity of the Secondary Flow

The effects of different line segment slopes on the average radial velocity of the inward flow and outward flow in the first turn were investigated, and the results are shown in Figure 11.

As illustrated in Figure 11, as the line segment slopes increase, the average radial velocities of the inward and outward flows corresponding to the adjusted region increase to varying degrees. This phenomenon contributes to the further acceleration of the radial migration of solid particles. Furthermore, the velocities of secondary flows in the stabilization phase were more significantly affected by the line segment slopes in regions I and III. Notably, when the line segment slope in region III increased from 6° to 10°, the increases in the average radial velocities of the inward flow and outward flow in this region reached 111.78% and 110.61%, respectively. Upon increasing the line segment slope in region I from 9° to 11°, the average velocity of the outward flow in this region exhibited an increase of 30.50%. In contrast, the secondary flow velocity of region II is relatively less affected by the adjustment to the line segment slope in the steady stage, resulting in a more uniform flow.

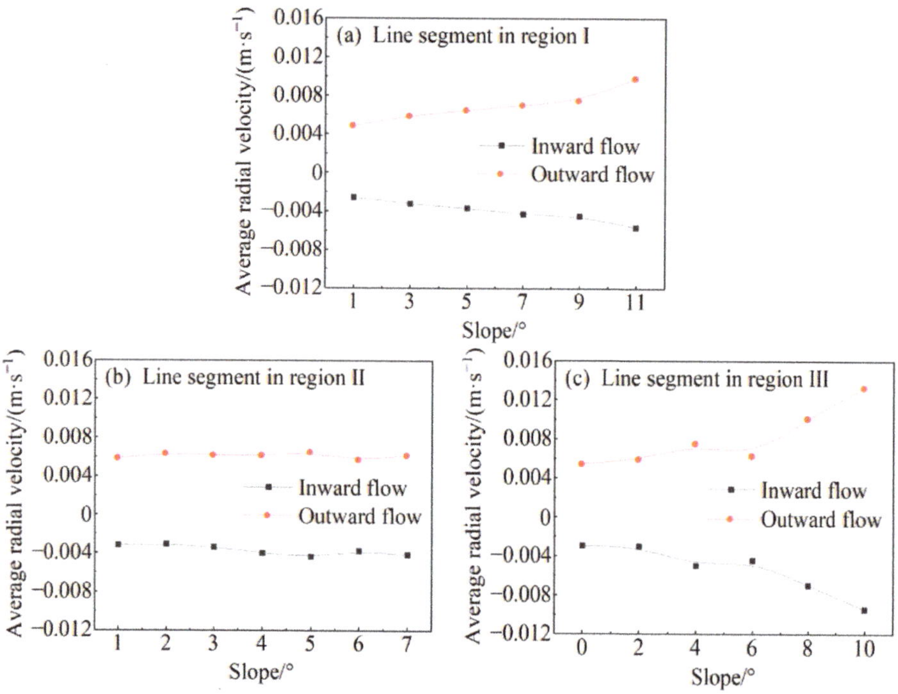

Figure 11. Effect of the line segment slopes in the localized region on the average radial velocity of the inward and outward flows in the third turn with different radial regions. (**a**) Region I; (**b**) region II; (**c**) region III.

3.2. *Effect of Localized Independent Adjustment of the Line Segment Slopes Within the Mixed Particle Zone on the Particle Separation Process*

3.2.1. Effect of Localized Independent Adjustment of the Line Segment Slopes Within the Mixed Particle Zone on the Radial Migration Behaviour of Particles

The radial motion of the fluid influences the force and radial migration motion of the particles. The effect of the line segment slope within the mixed particle zone on the radial migration behaviour of the particles is illustrated in Figures 12–14.

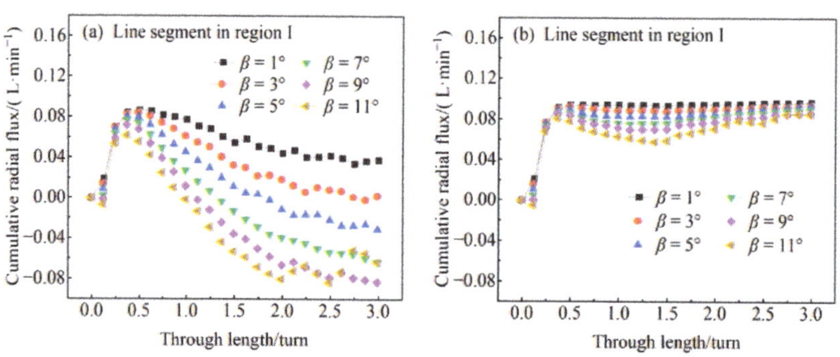

Figure 12. Effect of the line segment slope in region I on the cumulative radial flux at r = 80 mm. (**a**) Hematite; (**b**) quartz.

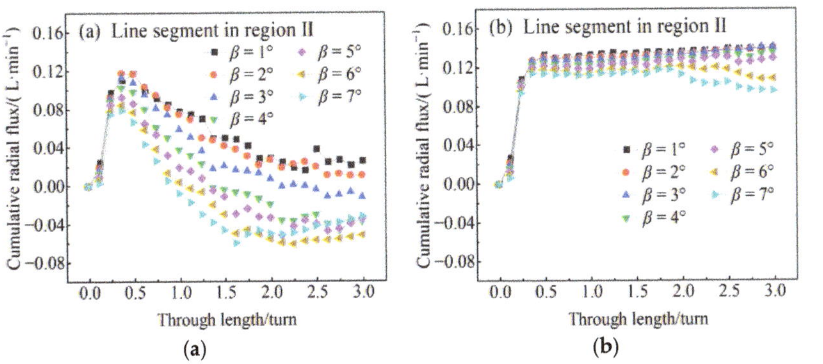

Figure 13. Effect of the line segment slope in region II on the cumulative radial flux at r = 100 mm. (**a**) Hematite; (**b**) quartz.

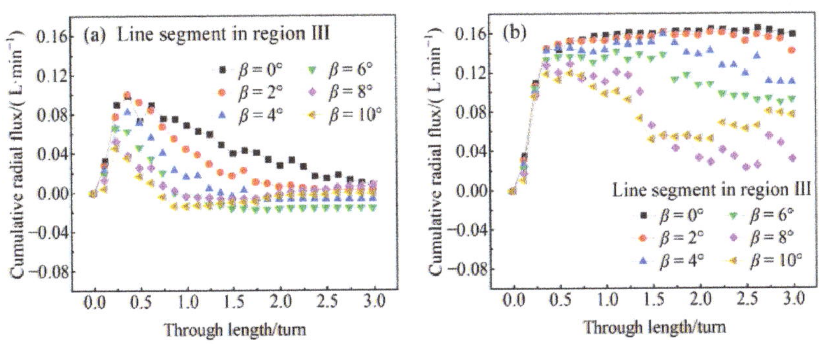

Figure 14. Effect of the line segment slope in region III on the cumulative radial flux at r = 120 mm. (**a**) Hematite; (**b**) quartz.

Figures 12a, 13a and 14a show that the cumulative radial fluxes of hematite particles are positive when the line segment slopes in regions I and II are low. This indicates that hematite particles migrate outwards, which is unfavorable for separation. Notably, the outward migration of hematite particles decreases with the gradual increase in the two line segment slopes. Upon increasing these two line segment slopes to 5° and 3°, the hematite particles exhibited a transition from outward to inward migration at a certain longitudinal travel of the fluid. The amount of inward migration increased with increasing longitudinal travel length and then stabilized. As the trough slope increases, the transition to inward motion becomes shorter and the cumulative radial flux increases. Nevertheless, when the line segment slope exceeds the regulatory range, there is a tendency for the inward migration of hematite in the steady state to decrease. Hematite particles in region III migrate predominantly outwards, and as the line segment slope in this region increases, the amount of outward migration gradually decreases, and the direction of particle migration transitions to inwards near the first turn and quickly reaches stabilization. Figures 12b, 13b and 14b indicate that the cumulative radial flux of quartz in regions I–III is positive, indicating that the migration direction is continuously outwards. When the line segment slope is fixed, the outward migration of quartz increases rapidly to a peak value in the first half-turn, after which it tends to decrease with increasing longitudinal travel. Moreover, the reduced magnitude of the outward migration of quartz is more pronounced from the inner to the outer regions of the trough, accompanied by minimal fluctuations and longer travel for the particle movement to stabilize in the radial direction. When the number of turns is kept constant, the outward migration of quartz in the trough is negatively correlated with the line segment slope. Furthermore, the impact of adjusting the line segment slopes within the mixed particle zone on the cumulative radial flux of quartz is more pronounced from the inner to the outer regions. Consequently, an increase in the line segment slopes in the regions I–III can induce the inward migration of hematite, which is conducive to improving the separation index of hematite and quartz particles. Nevertheless, the outward migration of quartz diminishes to varying degrees with increasing line segment slope, which also has a detrimental effect on the separation efficiency. Therefore, it is necessary to perform further calculations of the separation efficiency under different conditions to determine the appropriate range of adjustment for the line segment slope in each region.

3.2.2. Effect of Localized Independent Adjustment of the Line Segment Slopes Within the Mixed Particle Zone on the Separation Efficiency

Separation efficiency is one of the principal performance indicators for evaluating the separation performance of a spiral concentrator. The separation efficiencies of hematite and quartz at different splitter positions were calculated separately for independent adjustments of the localized line segment slope, with the results presented in Figure 15.

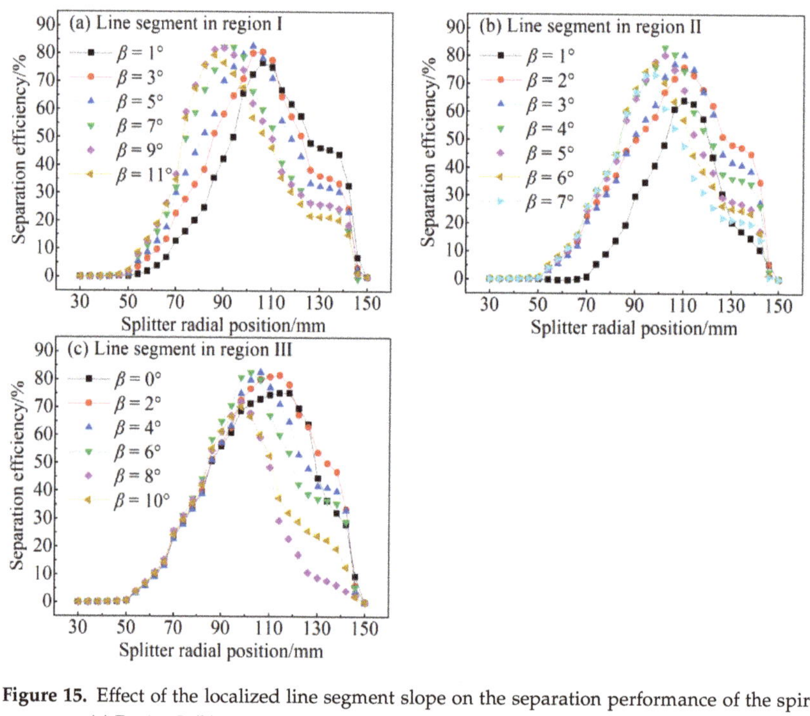

Figure 15. Effect of the localized line segment slope on the separation performance of the spiral separator. (**a**) Region I; (**b**) region II; (**c**) region III.

As illustrated in Figure 15, the maximum separation efficiencies of hematite and quartz initially tend to increase but then decrease immediately as the line segment slopes in regions I–III increase. Additionally, the radial position of the optimal interceptor is gradually shifted inwards. In the radial region inside the optimal splitter position, the line segment slopes in regions I and II exert a more pronounced influence on the enhancement of the separation efficiency. The ranges of the line segment slopes in each region with higher separation efficiencies are 5–9°, 3–5°, and 2–6°, respectively, and the maximum separation efficiencies are 81.94%–82.71%, 80.20%–83.01%, and 81.56%–82.58%, respectively.

3.3. Combination Adjustment of the Line Segment Slopes Within the Mixed Particle Zone and Structural Optimization

3.3.1. Effect of the Combination Adjustment of the Line Segment Slopes on the Maximum Separation Efficiency

The effects of the line segment slope within the mixed particle zone on the characteristics of the secondary flow and the particle separation process, as well as the range of each line segment slope used to achieve a relatively high separation efficiency, were subjected to a comprehensive analysis. In light of the above-described investigation, an optimization study was conducted on the combination mode of line segment slopes in accordance with the principles of Box–Behnken design. The line segment slope of region I (X_1), the line segment slope of region II (X_2), and the line segment slope of region III (X_3) were selected as the independent variables, with the separation efficiency E_s serving as the response value. The response surface factors and level settings are presented in Table 4, and the test levels of the independent variables are coded as −1, 0, and 1. The RSM results are presented in Table 5.

Table 4. Factors and levels of the Box–Behnken design.

Factors	Symbols	Levels		
		−1	**0**	**1**
Line segment slope of region I/°	X_1	1	6	11
Line segment slope of region II/°	X_2	1	4	7
Line segment slope of region III/°	X_3	0	4	8

Table 5. Results of the Box–Behnken design.

Source	Factors			Separation Efficiency/%
	X_1	X_2	X_3	
1	0	0	0	83.06
2	0	0	0	83.06
3	0	−1	1	72.84
4	1	1	0	73.71
5	1	0	−1	71.91
6	0	0	0	83.06
7	0	0	0	83.06
8	0	−1	−1	71.71
9	−1	0	−1	75.35
10	−1	0	1	67.68
11	0	0	0	83.06
12	0	1	1	74.69
13	1	−1	0	71.25
14	1	0	1	74.81
15	−1	1	0	69.45
16	−1	−1	−1	68.36
17	0	1	−1	77.50

The outcomes of the Box–Behnken design, as presented in Table 5, were subjected to a quadratic multiple regression analysis. This was performed on the basis of the maximum separation efficiencies achieved across various experimental conditions. A regression model was established between the line segment slope and the separation efficiency as follows:

$$Y = 83.86 + 1.35 \times X_1 + 1.4X_2 - 0.8062X_3 + 0.3425X_1X_2 + 2.64X_1X_3 - 0.985X_2X_3 - 7.46X_1{}^2 - 5.71X_2{}^2 - 3.97X_3{}^2$$

The correlation between the experimental values of the separation efficiency via numerical simulation and the values predicted via the regression model is shown in Figure 16. Additionally, the response surface multiple regression model was subjected to analysis of variance (ANOVA) and a significance test, as presented in Table 6.

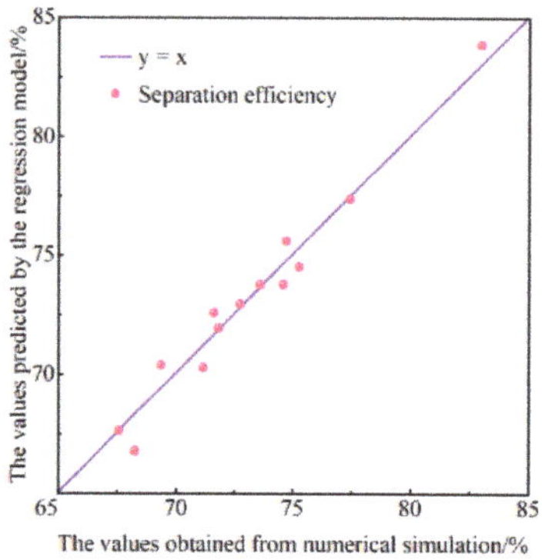

Figure 16. Correlations between the values obtained from numerical simulation and the values predicted by the regression model.

Table 6. Analysis of regression variance.

Source	Sum of Squares	Degrees of Freedom	Mean Square	F Value	p Value Prob > F	Degree of Significance
Model	490.4	9	54.49	79.78	<0.0001	**
X_1	14.69	1	14.69	21.51	0.0024	**
X_2	15.65	1	15.65	22.92	0.0020	**
X_3	5.20	1	5.20	7.61	0.0281	*
X_1X_2	0.4692	1	0.4692	0.6870	0.4345	
X_1X_3	27.93	1	27.93	40.90	0.0004	**
X_2X_3	3.88	1	3.88	5.68	0.0486	*
X_1^2	209.72	1	209.72	307.06	<0.0001	**
X_2^2	118.72	1	118.72	173.82	<0.0001	**
X_3^2	53.51	1	53.51	78.35	<0.0001	**
Residual	4.78	7	0.6830			
Lack of fit	4.78	3	1.59			
Pure error	0	4	0			
Total	495.19	16				

Notably, the coefficient of determination R^2 is 0.9903, whereas the corrected coefficient adjusted R^2_{Adj} is 0.9779. The following symbols are used to indicate statistical significance: *: $p < 0.05$, significant; **: $p < 0.01$, highly significant.

The results of ANOVA on the regression equation indicate that the response surface regression model presents a highly significant level of $p < 0.0001$, which can be utilized for subsequent optimization of the design using this response surface approximation model. The coefficient of determination, R^2, is 0.9903, and the correction coefficient, R^2Adj, is 0.9779, indicating that 99.03% of the variation in the response values is caused by the three factors examined. The regression model adequately describes the process by which the line segment slopes in three regions affect the separation efficiency. Figure 16 illustrates the correlation between the actual and predicted values of separation efficiency, further demonstrating the high credibility and accuracy of the test and the suitability of the fitted regression equation to the test principle.

The results of the significance test demonstrate that the primary terms X_1 and X_2, the interaction term X_1X_3, and the secondary terms X_{11}, X_{22}, and X_{33} have highly significant effects on the response value. Furthermore, the primary term X_3 and the interaction term

X_2X_3 significantly affect the response value. In contrast, there is no significant effect of other factors. This finding indicates that the line segment slopes of regions I and II have a highly significant effect on the separation efficiency, which is consistent with the aforementioned significant degree of effect on the separation efficiency when the line segment slope is adjusted independently, as well as the pattern of the change in the peak value of the radial fluxes of the inward flow in region I and the outward flow in region II at the initial stage (0.075th turn).

To intuitively reflect the strength of the interaction effect of the line segment slopes on the maximum separation efficiency, based on the results of the regression model analysis, the 3D response surface map and the 2D map of the contour projection when the three factors interact with each other two by two were further plotted, as shown in Figure 17. In accordance with the principles of response surface testing, a steeper response surface indicates a greater influence of the factors under test on the response value, and vice versa. The shape of the contour line reflects the strength of the interaction between any two factors. Furthermore, the shape of the contour line reflects the strength of the interaction between any two factors. Generally, a more elliptical contour line corresponds to a response value that is more sensitive to the interaction of the examined factors. Conversely, contour lines that are more rounded indicate a lower sensitivity to the interaction of the examined factors [39].

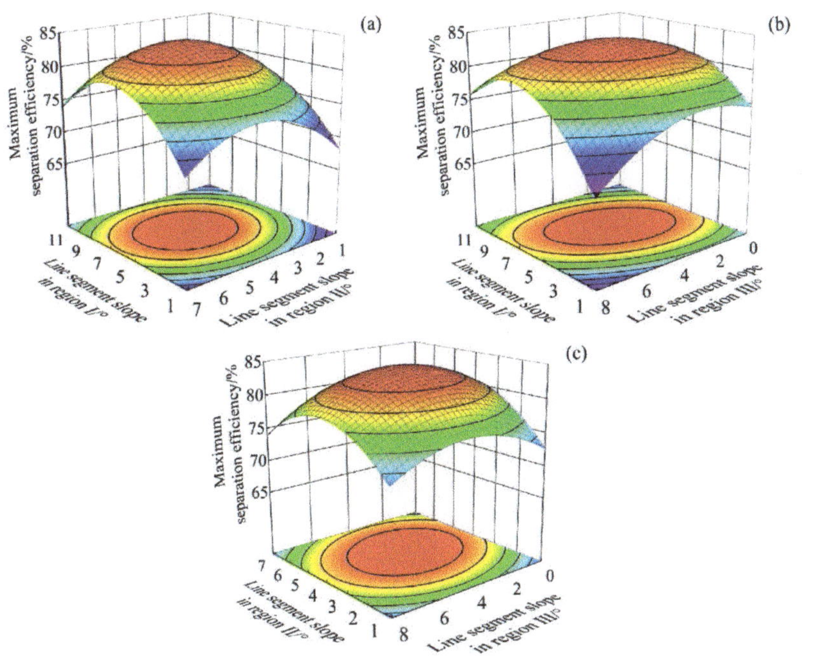

Figure 17. Three-dimensional response surface of the effects of the three factors on the maximum separation efficiency when the three factors interact in pairs. (**a**) Interaction of region I with region II; (**b**) interaction of region I with region III; (**c**) interaction of region II with region III.

As illustrated in Figure 17, the response surface is the most pronounced, and the eccentricity of the ellipse formed by the contour lines is greatest when regions I and III interact. This indicates that the synergistic effect of the line segment slopes in these two regions on the separation performance of the spiral concentrator is the most significant.

Further analysis of the regression model revealed that the optimal line segment slopes in the three regions were 6.40°, 4.43°, and 3.59°, respectively, at which point the predicted value of maximum separation efficiency was 83.26%, representing a 1.24 percentage point increase over the maximum separation efficiency (82.02%) achieved by the spiral concentrator with an optimal continuous curve section.

3.3.2. Verification of the Separation Performance of the Spiral Concentrator with an Optimal Combination of Line Segment Slopes

To verify the actual separation performance of the spiral concentrator after the combination adjustment of the line segment slopes, spiral concentrator test equipment with an optimal continuous curve section and an optimal combined line segment section were custom-machined. The photographs of the obtained test system are shown in Figure 18a, and the actual separation tests of hematite and quartz particles were carried out with the test system shown in Figure 18b. A series of numerical simulation tests and actual separation tests were conducted to assess the efficiency of the two types of spiral concentrators at varying splitter positions. The results are presented in Figure 19.

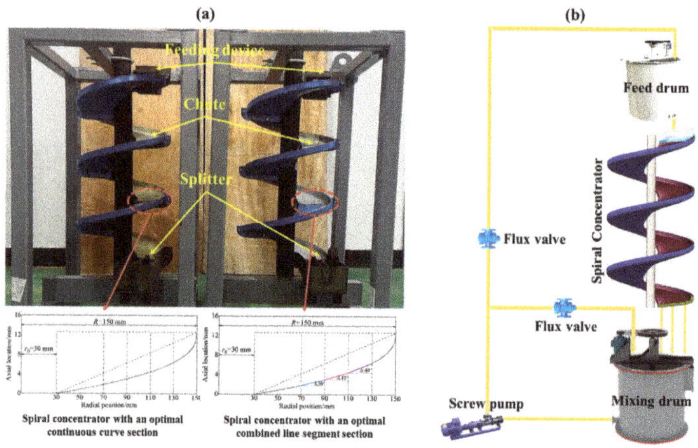

Figure 18. Spiral concentrator test equipment and separation test system. (**a**) Two optimized spiral concentrators; (**b**) systems for separation tests in spiral concentrators.

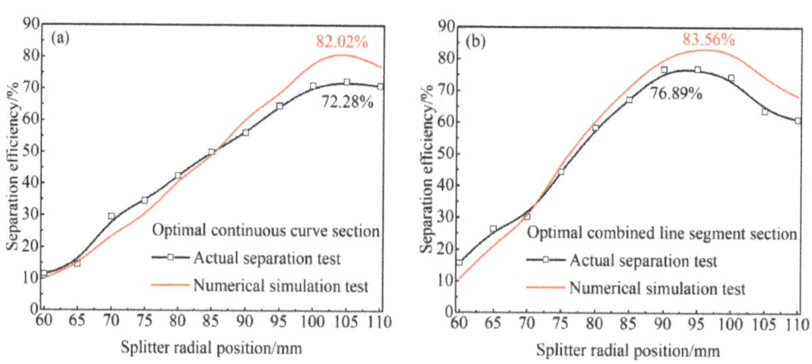

Figure 19. Comparison of the separation efficiency at various splitter positions for two types of spiral concentrators. (**a**) Optimal continuous curve section; (**b**) optimal combined line segment section.

As illustrated in Figure 19, the separation efficiencies obtained for the spiral concentrator with the optimal continuous curve section and optimal combined line segment section exhibit a trend identical to those predicted by the numerical simulation with regard to the outward movement of the splitter position. In both the actual separation tests and the numerical simulation tests, under identical structural parameters and feeding conditions, the maximum separation efficiency of the spiral concentrator with the optimal combined line segment section is greater than that of the spiral concentrator with the optimal continuous curve section. Furthermore, the peak shape of the curve is gentler, reducing the fluctuations in the production indices caused by adjusting the splitter position; this makes it easier to regulate and maintain the production indices. The above comparative analysis demonstrated that the actual separation efficiency of the spiral concentrator for hematite and quartz can be increased from 72.28% to 76.89% through the optimal configuration of the cross-sectional geometry within the mixed particle zone; this represents a significant improvement in the separation performance of the optimized equipment, and verifies the reliability and high efficiency of the numerical models and the optimal design methods for section geometry.

The numerical simulation results of the spiral concentrator have more or less deviations from the actual test results. For example, the neglected roughness of the trough, the simplified particle size composition, and other factors will affect the evolution of the flow field characteristics in the spiral concentrator, which will affect the particles' movement behaviour accordingly. So, the separation efficiency obtained from numerical simulation is slightly better than that of the actual separation experiment.

3.3.3. Comparison of the Radial Migration Behaviour of Particles Before and After Combination Adjustment for the Line Segment Slopes

To analyze the theoretical mechanism by which the localized adjustment of the line segment slopes within the mixed particle zone of the spiral concentrator affects the radial migration behaviour of the particles, the variation in the cumulative radial fluxes of the particles at the three representative radial positions was examined. The results are shown in Figure 20.

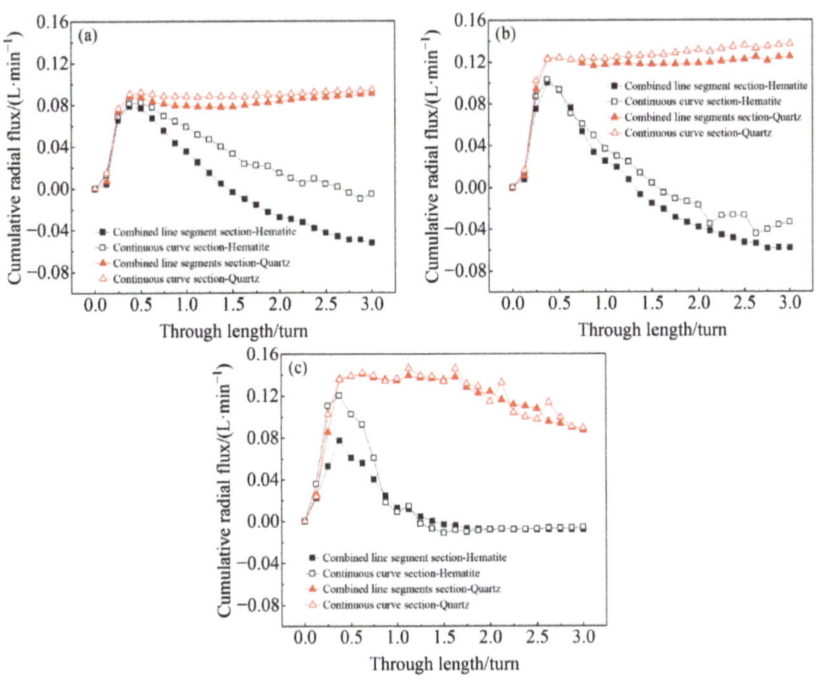

Figure 20. Variation curves of the cumulative radial flux of particles in spiral concentrators with different cross-sections. (**a**) $r = 80$ mm; (**b**) $r = 100$ mm; (**c**) $r = 120$ mm.

Figure 20 illustrates that following the combination adjustment of the cross-sectional slope within the mixed particle zone; there is a notable increase in the inward migration of hematite in regions I and II, whereas the outward migration of quartz is reduced. This indicates that the amount of stagnant quartz increases. Nevertheless, the increase in the inward migration of hematite is significantly greater than that of quartz, and the difference in the inward migration between the two is increased, which is conducive to improving the separation efficiency of hematite and quartz. The radial migration of quartz in region III varies relatively little, whereas the outward migration of hematite decreases significantly in the first turn. This phenomenon enhances the separation efficiency at the initial stage of flow-field evolution and shortens the travel distance for the separation process.

4. Conclusions

This paper examines the effects of adjusting the cross-sectional slope within the mixed particle zone on the secondary flow characteristics and particle separation process in a spiral concentrator. Hematite and quartz particles with a narrow size range (−100 + 75 μm for the practical separation experiment and 89.5 μm for the numerical simulation) were chosen for the separating objects based on the particle size composition characteristics of the feed of spiral concentrators in a processing plant of China. The independent and interactive effects of partial slopes on the spiral separation performance are evaluated systematically. The main findings can be summarized as follows:

(1) The peak radial flux of the secondary flow in each partition of the mixed particle zone is positively correlated with the slope of each line segment. The slopes of the line segment in region I (r = 70–90 mm) and that in region II (r = 90–110 mm) significantly affect the fluxes of the inward and outward flows, respectively.
(2) When the flow field in the trough becomes stable, the average radial velocity of the secondary flow in each partition of the mixed particle zone is also positively correlated with the slope of each line segment. The slopes of the line segment in regions I (r = 70–90 mm) and III (r = 110–130 mm) strongly influence the radial velocity.
(3) Increasing the slope of each line segment promotes the inward migration motion of hematite particles and reduces the outward migration of quartz. An increase in the line segment slopes of regions I to III to ranges of 5–9°, 3–5°, and 2–6°, respectively, contributes to the higher efficiency of the separation of the hematite and quartz particles, which is consistent with their migration variation.
(4) The independent adjustment of the slopes in regions I and II has a significant effect on the separation efficiency and is highly correlated with the radial flux changes in the secondary flow. The slopes in regions I and III were identified as the most significant combination of interaction effects. The optimal slopes in regions I to III were 6.40°, 4.43°, and 3.59°, respectively.
(5) The actual separation efficiency for hematite and quartz particles with a size range of −100 + 75 μm was increased from 72.28% to 76.89% by using the experimental spiral concentrator with the cross-sectional profile of optimal combined line segments that benefited from the increased inward migration difference between hematite and quartz particles and the decreased outward migration of hematite at the initial stage of the flow.

Author Contributions: S.G.: Conceptualization, methodology, writing—review and editing, project administration, resources, data curation. Q.W.: Writing—original draft and editing, software, data curation, formal analysis, visualization. X.Z.: Resources, investigation, software, validation. C.L.: Resources, investigation, software, validation. Y.S.: Supervision, writing—review, project administration, funding acquisition. B.C.: Supervision, writing—review, project administration, funding acquisition. All authors have read and agreed to the published version of the manuscript.

Funding: This research was funded by the [National Natural Science Foundation of China] grant number [52274257; 51974065], the [Project of Zhongyuan Critical Metals Laboratory] grant number [GJJSKFYB202406], and by the [Fundamental Research Funds for the Central Universities] grant number [N2201008; N2201004].

Data Availability Statement: Data will be made available on request.

Acknowledgments: The authors gratefully acknowledge financial support from the National Natural Science Foundation of China (Grant No. 52274257; 51974065), the Project of Zhongyuan Critical Metals Laboratory (Grant No. GJJSKFYB202406), and the Fundamental Research Funds for the Central Universities (Grant No. N2201008; N2201004).

Conflicts of Interest: The authors declare that they have no known competing financial interests or personal relationships that could have appeared to influence the work reported in this paper.

eferences

de Freitas Marques Araujo, P.; Silva, P.M.P.; Carmo, A.L.V.D.; Goncalves, M.V.L.; Costa, R.V.D.; de Melo, C.C.A.; Lucheta, A.R.; Montini, M. Gravity methods applied to bauxite residue for mineral pre-concentration. In *Light Metals 2021*, 50th ed.; Springer: Berlin/Heidelberg, Germany, 2021; pp. 68–76. [CrossRef]

Dixit, P.; Tiwari, R.; Mukherjee, A.K.; Banerjee, P.K. Application of response surface methodology for modeling and optimization of spiral separator for processing of iron ore slime. *Powder Technol.* **2015**, *275*, 105–112. [CrossRef]

Kademli, M.; Gulsoy, O.Y. The role of particle size and solid contents of feed on mica-feldspar separation in gravity concentration. *Physicochem. Probl. Miner. Process* **2012**, *48*, 645–654. [CrossRef]

Khanchi, A.R.; Sedighi, H.; Ansar, S.; Fasihi, J. Preconcentration of rare earth elements from iranian monazite ore by spiral separator using multi-response optimization method. *Int. J. Min. Sci. Technol.* **2014**, *24*, 117–121. [CrossRef]

Sadeghi, M.; Bazin, C.; Renaud, M. Radial distribution of iron oxide and silica particles in the reject flow of a spiral concentrator. *Int. J. Miner. Process.* **2016**, *153*, 51–59. [CrossRef]

Yu, W.; Zhang, H.L.; Wang, X.B.; Rahman, Z.U.; Shi, Z.C.; Bai, Y.H.; Wang, G.H.; Chen, Y.Q.; Wang, J.J.; Liu, L.J. Enrichment of residual carbon from coal gasification fine slag by spiral separator. *J. Environ. Manag.* **2022**, *315*, 115149. [CrossRef]

Bahrami, A.; Farajzadeh, S.; Kazemi, F. The effect of particle size distribution and liberation degree on the separation performance of industrial spirals in low-grade chromite processing. *Min. Metall. Explor.* **2021**, *38*, 277–287. [CrossRef]

Kapur, P.C.; Meloy, T.P. Industrial modeling of spirals for optimal configuration and design: Spiral geometry, fluid flow and forces on particles. *Powder Technol.* **1999**, *102*, 244–252. [CrossRef]

Jain, P.K. A novel concept of composite coefficient of friction for slurry systems and development of flow equations for spiral concentrators. *Powder Technol.* **2021**, *386*, 411–419. [CrossRef]

Gao, S.L.; Zhou, X.H.; Meng, L.G.; Zhao, Q.; Liu, W.G. Variation of flow hydrodynamic parameters and prediction of particle separation indices in the spiral concentrator with the regulation of pitch-diameter ratio. *Separations* **2023**, *10*, 410. [CrossRef]

Holland-Batt, A.B. Some design considerations for spiral separators. *Miner. Eng.* **1995**, *8*, 1381–1395. [CrossRef]

Davies, P.O.J.; Goodman, R.H.; Deschamps, J.A. Recent developments in spiral design, construction and application. *Miner. Eng.* **1991**, *4*, 437–456. [CrossRef]

Szpyrka, J.; Suponik, T.; Lutynski, M. Impact of spiral separator geometrical parameters on the density separation of various fine-grained materials. *Miner. Eng. Conf. (MEC2017)* **2017**, *18*, 01035. [CrossRef]

Ye, G.C.; Ma, L.Q.; Alberini, F.; Xu, Q.; Huang, G.; Yu, Y.X. Numerical studies of the effects of design parameters on flow fields in spiral concentrators. *Int. J. Coal Prep. Util.* **2022**, *42*, 67–81. [CrossRef]

Tripathy, S.K.; Murthy, Y.R. Modeling and optimization of spiral concentrator for separation of ultrafine chromite. *Powder Technol.* **2012**, *221*, 387–394. [CrossRef]

Ye, G.C.; Huo, Y.Y.; Li, C.F.; Deng, C.; Yu, Y.X.; Huang, G.; Ma, L.Q. A comparative study of trough profile and operating parameters performance in spiral concentrator. *Int. J. Coal Prep. Util.* **2021**, *41*, 678–691. [CrossRef]

Arnold, D.J.; Stokes, Y.M.; Green, J.E. Thin-film flow in helically-wound rectangular channels of arbitrary torsion and curvature. *J. Fluid Mech.* **2015**, *764*, 76–94. [CrossRef]

Nienaber, E.C.; Auret, L. Spiral concentrator interface monitoring through image processing: Optimization for parameter selection. *IFAC-PapersOnLine* **2016**, *49*, 102–107. [CrossRef]

Gao, S.L.; Meng, L.G.; Zhou, X.H.; Shen, Y.B.; Cui, B.Y.; Song, Z.G. Design of partial cross-sectional geometry and prediction of separation performance in the spiral separator. *Sep. Sci. Technol.* **2022**, *57*, 2127–2144. [CrossRef]

Kwon, J.; Kim, H.; Lee, S.; Cho, H. Simulation of particle-laden flow in a humphrey spiral concentrator using dust-liquid smoothed particle hydrodynamics. *Adv. Powder Technol.* **2017**, *28*, 2694–2705. [CrossRef]

21. Meng, L.G.; Gao, S.L.; Wei, D.Z.; Cui, B.Y.; Shen, Y.B.; Song, Z.G.; Yuan, J. Effects of cross-sectional geometry on flow characteristi in spiral separators. *Sep. Sci. Technol.* **2021**, *56*, 2967–2977. [CrossRef]
22. Boucher, D.; Deng, Z.; Leadbeater, T.; Langlois, R.; Renaud, M.; Waters, K.E. PEPT studies of heavy particle flow within a spir concentrator. *Miner. Eng.* **2014**, *62*, 120–128. [CrossRef]
23. Boucher, D.; Deng, Z.; Leadbeater, T.; Langlois, R.; Waters, K.E. Observation of iron ore beneficiation within a spiral concentrat by positron emission particle tracking of large (Φ = 1440 μm) and small (Φ = 58 μm) hematite and quartz tracers. *Chem. Eng. S* **2016**, *140*, 217–232. [CrossRef]
24. Boucher, D.; Deng, Z.; Leadbeater, T.W.; Langlois, R.; Waters, K.E. Speed analysis of quartz and hematite particles in a spir concentrator by PEPT. *Miner. Eng.* **2016**, *91*, 86–91. [CrossRef]
25. Matthews, B.W.; Fletcher, C.; Partridge, A.C. Computational simulation of fluid and dilute particulate flows on spiral concentrato *Appl. Math. Model.* **1998**, *22*, 965–979. [CrossRef]
26. Matthews, B.W.; Fletcher, C.; Partridge, A.C.; Vasquez, S. Computations of curved free surface water flow on spiral concentrato *J. Hydraul. Eng.-ASCE* **1999**, *125*, 1126–1139. [CrossRef]
27. Mishra, B.K.; Tripathy, A. A preliminary study of particle separation in spiral concentrators using DEM. *Int. J. Miner. Proce.* **2010**, *94*, 192–195. [CrossRef]
28. Doheim, M.A.; Gawad, A.F.A.; Mahran, G.M.A.; Abu-Ali, M.H.; Rizk, A.M. Numerical simulation of particulate-flow in spir separators: Part I. Low solids concentration (0.3% & 3% solids). *Appl. Math. Model.* **2013**, *37*, 198–215. [CrossRef]
29. Rong, W.J.; Feng, Y.Q.; Schwarz, P.; Witt, P.; Li, B.K.; Song, T.; Zhou, J.W. Numerical study of the solid flow behavior in a rotatir drum based on a multiphase CFD model accounting for solid frictional viscosity and wall friction. *Powder Technol.* **2020**, *36* 87–98. [CrossRef]
30. Ferreira, S.L.C.; Bruns, R.E.; Ferreira, H.S.; Matos, G.D.; David, J.M.; Brandão, G.C.; Da Silva, E.G.P.; Portugal, L.A.; Reis, P.S.I Souza, A.S.; et al. Box-behnken design: An alternative for the optimization of analytical methods. *Anal. Chim. Acta* **2007**, *59* 179–186. [CrossRef]
31. Meng, L.G.; Gao, S.L.; Wei, D.Z.; Zhao, Q.; Cui, B.Y.; Shen, Y.B.; Song, Z.G. Particulate flow modelling in a spiral separator b using the Eulerian multi-fluid VOF approach. *Int. J. Min. Sci. Technol.* **2023**, *33*, 251–263. [CrossRef]
32. Syamlal, M.; Rogers, W.; Brien, T.J.O. *MFIX Documentation Theory Guide*; USDOE Morgantown Energy Technology Center (METC Morgantown, WV, USA, 1993. [CrossRef]
33. Hou, D.X.; Cui, B.Y.; Zhao, Q.; Wei, D.Z.; Song, Z.G.; Feng, Y.Q. Research on the structure of the cylindrical hydrocyclone spig to mitigate the misplacement of particles. *Powder Technol.* **2021**, *387*, 61–71. [CrossRef]
34. Jain, P.K.; Rayasam, V. An analytical approach to explain the generation of secondary circulation in spiral concentrators. *Powd Technol.* **2017**, *308*, 165–177. [CrossRef]
35. Gidaspow, D.; Jung, J.; Singh, R.K. Hydrodynamics of fluidization using kinetic theory: An emerging paradigm: 2002 Flour-Dani lecture. *Powder Technol.* **2004**, *148*, 123–141. [CrossRef]
36. Ng, B.H.; Ding, Y.L.; Ghadiri, M. Assessment of the kinetic-frictional model for dense granular flow. *Particuology* **2008**, *6*, 50–5 [CrossRef]
37. Launder, B.E.; Spalding, D.B. The numerical computation of turbulent flows. *Comput. Meth. Appl. Mech. Eng.* **1974**, *3*, 269–28 [CrossRef]
38. Schulz, N.F. Separation efficiency. *Trans. Soc. Min. Eng. AIME* **1970**, *247*, 81–87. Available online: https://cir.nii.ac.jp/crid/1570 2701370836096 (accessed on 7 December 2024).
39. Hill, W.J.; Hunter, W.G. A review of response surface methodology: A literature survey. *Technometrics* **1966**, *8*, 571–590. [CrossRe

minerals

MDPI

rticle

Study on the Preparation of High-Quality Quartz and Its Mechanism by Combining Pretreatment with Metallurgy

ongjun Huang * and Ning Zhang

School of Minerals Processing and Bioengineering, Central South University, Changsha 410083, China; 225611052@csu.edu.cn
* Correspondence: 207049@csu.edu.cn

Abstract: As a strategically important resource with stable properties, high-purity quartz plays a crucial role in various fields such as high-purity quartz crucibles, semiconductors, and electronics. Currently, the availability of high-quality quartz resources is gradually diminishing, and there are no high-quality quartz deposits in some areas, making the improvement of low-quality quartz essential. This study focuses on the quartz sand produced through river dredging in Yueyang. The detected SiO_2 content is 92.31%, which enables the secondary utilization of waste resources and provides significant environmental benefits. In this study, the iron oxide removal efficiency achieved through magnetic separation pretreatment can reach 87.24%. Following flotation, calcination, and leaching processes, high-purity quartz products with a total impurity content of less than 100 ppm were successfully obtained. In addition, the factors affecting magnetic separation and calcination effect were studied, and the process parameters for reference were obtained. The mechanism of microwave impurity removal was investigated using SEM, TG-MS, and in-situ XRD, demonstrating the feasibility of producing high-purity quartz from low-quality quartz.

Keywords: high-purity quartz; magnetic separation; microwave calcination; river sand

1. Introduction

High-purity quartz is utilized in the aerospace, semiconductor, photovoltaic, electronics, and other emerging industries due to its excellent physical and chemical properties [1]. In recent years, with the development of high-tech industries, high-purity quartz production is in short supply, and quartz resources with higher purity are being gradually depleted. Moreover, some countries do not have high-purity quartz deposits at all. It is very urgent to develop the technology of using low-quality quartz resources to prepare high-purity quartz. Although the low-quality quartz resources such as powder quartz, quartz sandstone and river sand have huge reserves, they are generally used only in the low-end ceramics and metallurgy industries [2]. If we can improve their purity through better technology, there will be more raw materials for high-tech industries.

Depending on their different forms, impurities in quartz can be divided into gangue impurities, inclusions, and impurities in the form of lattice substitution with minerals. Among them, the independent gangue impurities in quartz minerals, such as mica, rutile, feldspar, iron ore, clay mineral and other impurities, can be identified and analyzed by PM (microprobe), EMP (electron microprobe), EDS (Energy Dispersive Spectrometer) and other means [3]. Most of these impurities can be removed by color separation, gravity separation, magnetic separation, flotation and other mineral processing means [4–6]. Among these impurities, iron oxide is the most persistent in quartz ore, and its concentration determines the usage of quartz in various industries. This is because iron oxide significantly affects the color and performance of quartz products. If the iron content in quartz is excessively high, it will diminish the whiteness of ceramic products, lower the melting point of refractories, reduce the transparency of glass, and decrease the transmission efficiency of semiconductor materials [7]. The methodology for eliminating iron-containing impurities from quartz through magnetic separation has reached a relatively advanced stage. For instance, the research conducted by Sekulic et al. [8]. identified that the state of occurrence of iron impurities significantly influences the removal efficiency of iron oxides. Additionally, researchers including Li and colleagues employed superconducting high-gradient magnetic separation techniques to investigate the extraction of minerals from sea sand. In this process, while achieving a concentration of sea sand, they also attained an impressive Fe element removal rate of 82.10% [5].

The fluid inclusion impurities in quartz sand, which consist of liquid and gas trapped within the quartz crystal during the mineralization process, primarily include H_2, CO_2, O_2, N_2, CH_4, CO and other impurities. These can be detected using FTIR (Fourier Transform Infrared Spectroscopy), La-RS (Laser Raman Spectroscopy), EMP analysis and other methods [9–11]. In fluid inclusions, the most important is the hydroxyl content of water molecules, followed by carbon-containing compounds such as CO_2, CO, and hydrocarbon fluid impurities. B. G. et al. showed that the large gas–liquid inclusions in quartz could be broken by low temperature roasting, thus the inclusions inside the lattice were volatilized and exposed [12]. In addition, in the process of mineralization, Al^{3+}, Fe^{3+}, Ti^{2+} and other elements will enter the quartz through the form of lattice substitution and become solid inclusions that are difficult to remove. K^+, Na^+, Ca^{2+} will also exist in the quartz lattice in the form of compensation ions or fluid inclusions [11,13], which are difficult to remove by traditional mineral processing methods.

Calcination and leaching are mature and effective methods in purifying high-purity quartz at present. To further enhance the removal of impurities, researchers have developed advanced calcination techniques involving ultra-high temperatures, vacuum and dopant methods, as well as improved leaching methods using ultrasonic, hot pressing, pre-alkali corrosion and other means [14–19]. Conventional high-temperature calcination often takes a long holding time and requires a lot of energy consumption. Therefore, it is necessary to optimize the calcination method, such as through microwave calcination for quartz pretreatment. Microwave calcination can make the sample more evenly heated, and reduce the calcination time and energy consumption significantly [20]. Furthermore, Yuan et al. investigated the calcination of potassium feldspar using calcium chloride and observed that the addition of chlorine salt significantly lowered the calcination temperature of potassium feldspar [21]. Given that the removal of calcium chloride is challenging and could potentially impact the quality of high-purity quartz, NH_4Cl, NaCl, or KCl is employed as the chlorination agent during the roasting process to expedite the roasting effect of potassium feldspar.

Consulting the literature, most of the studies in this direction are based on the analysis of the quartz after roasting, and research on the mechanism of the heating process of roasting is lacking. The objective of this study is to conduct a comprehensive analysis of the process mineralogy and inclusion impurities present in river sand from Yueyang, with the aim of identifying the primary impurities and their states of occurrence within the ore. The most significant iron impurities that influence the properties of quartz were eliminated through magnetic separation pretreatment, while also investigating the factors that affect its efficacy. Subsequently, flotation, calcination, and leaching processes were employed to examine variables impacting calcination effectiveness, thereby facilitating the removal of additional impurities from quartz. This research further explores the feasibility of producing high-purity quartz from low-quality sources. Additionally, the TG-MS and in-situ XRD detection techniques utilized in this study effectively capture the variations in gas composition and crystal types during the calcination heating process, thus establishing a theoretical foundation for the exploitation of similar low-quality quartz sand.

2. Materials and Methods

2.1. Raw Materials

The raw ore used in this experiment is a kind of river sand in Yueyang County, Hunan Province. The particle size distribution of the raw ore obtained through screening analysis is presented in Table 1, indicating a particle size range of 0.05 mm to 5 mm for the raw ore. The surface of the minerals can be obviously seen with mud impurities. After the raw ore is mixed and the sample is prepared, XRF chemical full-element analysis is performed, and the elements with a higher content that can represent the types of impurities are listed, as shown in Table 2. The results show that the proportion of SiO_2 in the raw ore is 92.31%, while the impurities with higher content are Al_2O_3 and K_2O. In addition, the Fe_2O_3 content in the raw ore is 0.17%, which will affect the application of quartz products in various fields. The reagents selected in this experiment are AR (Analytical Reagent, Shanghai Macklin Biochemical Technology Co., Ltd., Shanghai, China), and the conductivity of ultra-pure water used in the experiment is 18.2 MΩ·cm.

Table 1. Size distribution of raw ore.

Particle Size/mm	Content/%
+4.75	2.03
4.75~2	14.15
2~1	27.37
1~0.45	31.21
0.45~0.15	22.86
0.15~0.074	1.77
0.074~0.05	0.20
−0.05	0.40

Table 2. Main chemical elements of raw ore.

Element	Content (wt%)
SiO_2	92.31
Al_2O_3	2.42
K_2O	1.76
Na_2O	0.08
MgO	0.02
Fe_2O_3	0.17
CaO	0.06
TiO_2	0.04

2.2. Purification Process for Quartz Samples

First use the ball mill to grind the raw ore (DH48S horizontal ball mill adopts zirconia ball medium and ceramic cavity). Then the fine quartz product after grinding is added to the agitator for scrubbing and desliming to obtain clean fine quartz. The fine-grained quartz underwent treatment with a high-gradient magnetic separator (DLSD type, Yueyang Hercules Electromagnetic Machinery, Yueyang, China), and the experiment of magnetic separation conditions was carried out by using the method of controlling variables.

Subsequently, the best non-magnetic products were reversed flotation by XFD-IV (Changchun Prospecting Machinery Factory, Changchun, China) type single cell flotation machine. The overall beneficiation flow chart is depicted in Figure 1. The ultimate objective is to obtain quartz fine powder with the proportion of SiO_2 exceeding 99.6%, which will serve as raw material for subsequent calcination condition experiments and mechanism research.

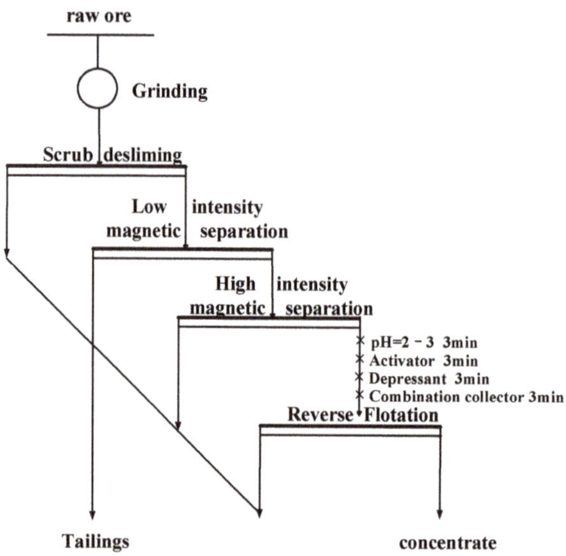

Figure 1. Beneficiation flowsheet.

The above obtained quartz products were used as raw materials, and the calcination conditions were tested using a microwave tube furnace (CY-HT1600C-M, Changyi Microwave Technology Company, Changyi, China) at a frequency of 2.45 Ghz. The 18 g of quartz and 2 g of calcined dopant were ground using a three-head grinder for 1 min to ensure even mixing. The mixed ore was then placed into a quartz boat and inserted into the microwave tube furnace for the calcination condition test. Afterward, the high-temperature quartz was immediately placed into cold pure water to cool down, washed with pure water 5 times to remove surface salt, dried, and then subjected to leaching using a mixed solution of HF, HCl, and HNO_3 (0.5 mol/L, 3 mol/L, 1 mol/L) at 80 °C for 4 h, with a liquid–solid ratio of 10:1. Following the leaching process, the samples were cleaned with pure water 5 times and dried in an drying oven at 85 °C for 24 h. The obtained samples were ground to 38 μm using a three-head grinder, dissolved, and analyzed using ICP-OES. The experimental setup is illustrated in Figure 2.

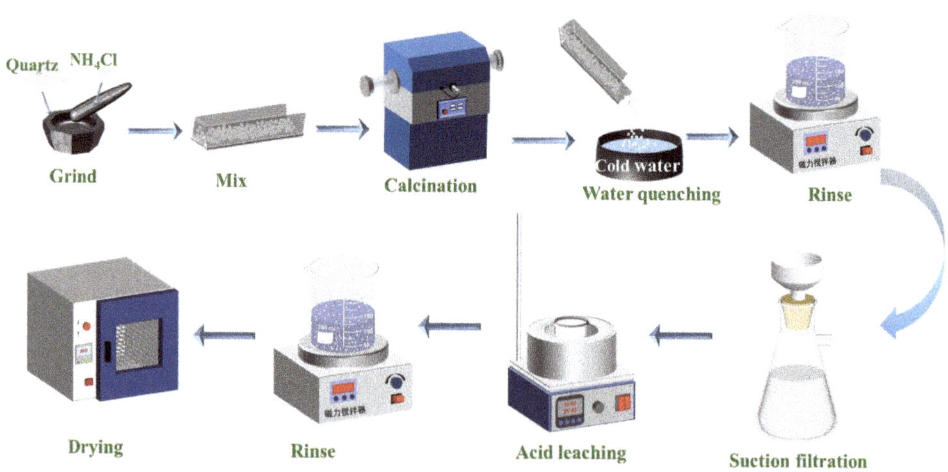

Figure 2. Calcination experiment flowsheet.

All experiments in this paper were conducted in parallel three times, and the data in all tables are the results after averaging.

The product yield, recovery rate and removal rate are calculated by the following formula:

$$\gamma_a = \frac{M_p}{M_f} \quad (1)$$

where γ_a is the yield of product a, M_p is the quality of product a, and M_f is the quality of feed ore;

$$\varepsilon_{ai} = \frac{\gamma_a \beta_{ai}}{\beta_{fi}} \quad (2)$$

where ε_{ai} is recovery rate of a certain composition in a product, γ_a is yield of the product, β_{ai} is content of the certain composition in the product, β_{fi} is content of the certain composition in the feed.

$$R_{ai} = 1 - \varepsilon_{ai} \quad (3)$$

where R_{ai} is removal rate of a certain composition in a product, γ_a is yield of the product.

2.3. X-Ray Diffraction (XRD) Analysis

The sample to be tested was mixed and divided, and ground with a three-head grinder to less than 38 μm, which was pressed into sheets for XRD analysis with an X-ray diffractometer (X'Pert3 Powder, Malvern Panalytical Co., Ltd., Almelo, The Netherlands). The X-ray light tube was the Cu target. The operating voltage of the equipment was set to 40 kV and operating current was set to 40 mA. The θ/θ scanning mode was adopted. The scanning range was set from 3° to 80°.

2.4. Optical Microscope Analysis

In order to eliminate the interference of surface impurities and further study the internal inclusions of minerals, the minerals after flotation were observed with an optical microscope: place a small number of mineral particles on the surface of the slide, slowly drop two drops of paraffin oil until it is completely submerged, drive away the bubbles. Another slide was covered and the inclusions were observed with a polarizing microscope (Leica-DM RXE, Leica GMBH, Wetzlar, Germany).

2.5. Raman Spectral Analysis

The pre-treated quartz was ground to less than 74 μm, and a small amount of samples were taken and placed on a slide to flatten. We used a micro-confocal Raman spectrometer (Renishaw Invia Qontor, Renishaw (Hong Kong) Limited, Hong Kong) to observe the internal inclusions of minerals. The excitation wavelength was 532 nm and the scanning time was 10 s.

2.6. In-Situ X-Ray Diffraction Analysis

In-situ XRD (Panalytical Empyrean) is used to analyze changes in the roasting process of minerals, and the temperature measurement range is from 400 °C–1100 °C. The scanning rate is 20°/min, and the holding time for each temperature point is 5 min. The X-ray light tube was the Cu target. The operating voltage of the equipment was set to 40 kV and operating current was set to 40 mA. The θ/θ scanning mode was adopted. Scanning range was set from 3° to 140°.

2.7. Scanning Electron Microscope (SEM) Analysis

The samples were sprayed with gold, and the quartz impurity distribution and surface morphology were determined by scanning electron microscopy (TESCAN MIRA LMS,

Kohoutovice, Czech Republic). The working voltage was set at 15 keV; the detection distance was 10 mm; the acceleration current was 11 mA.

2.8. Inductively Coupled Plasma-Optical Emission Spectrum (ICP-OES) Analysis

The trace element content of quartz concentrate was determined by ICP-OES (SPECTRO BLUE SOP, SPECTRO Analytical Instruments, Kleve, Germany). The RF generator frequency was 27.12 MHz, the wavelength range was 165–770 nm, and the UV photo sealing argon-filled cycle system was adopted.

2.9. Thermogravimetric Mass Spectrometry (TG-MS) Analysis

The pre-treated quartz sample was evenly mixed with chlorination agent at a ratio of 100:2, and the composition changes during calcination were analyzed by TG-MS (Netzsch STA449F3, Netzsch QMS403Q, Netzsch, Selb, Germany). The temperature measurement range was from 30 °C to 1130 °C, and the heating rate was 20 °C/min.

2.10. X-Ray Fluorescence Spectrum (XRF) Analysis

XRF can determine elements from oxygen to uranium. An X-ray fluorescence spectrometer (Axios Max, Malvern Panalytical Co., Ltd., Almelo, The Netherlands) was used for XRF analysis. XRF samples preparation was similar to XRD. The equipment adopted a SST-mAX non-tungsten filament X-ray tube. The operating voltage and current were set to 20 kV and 10 mA, respectively.

3. Results and Discussion

3.1. Characteristics of Raw Quartz

XRD analysis of the raw ore (Figure 3) shows that the main mineral in the raw ore is quartz, followed by feldspar, and no other obvious impurity phase has been detected. Combined with chemical element analysis, it can be seen that the main impurity mineral is potassium feldspar, so the separation of feldspar and quartz minerals is mainly carried out. Figure 4 is the SEM image of quartz sand raw ore, in which quartz and feldspar minerals are closely related to each other. As can be seen in Figure 4d, most of the raw ore is independent mineral of quartz or feldspar, but there are still some connected biotite particles that need to be separated by grinding, and there are inclusions in quartz minerals which need to be further confirmed.

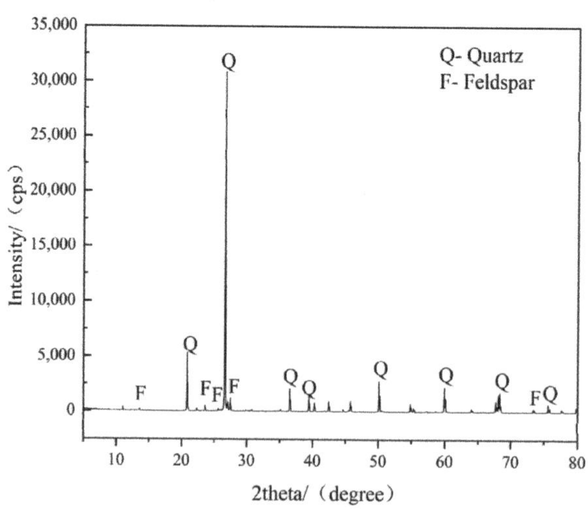

Figure 3. XRD pattern of raw ore.

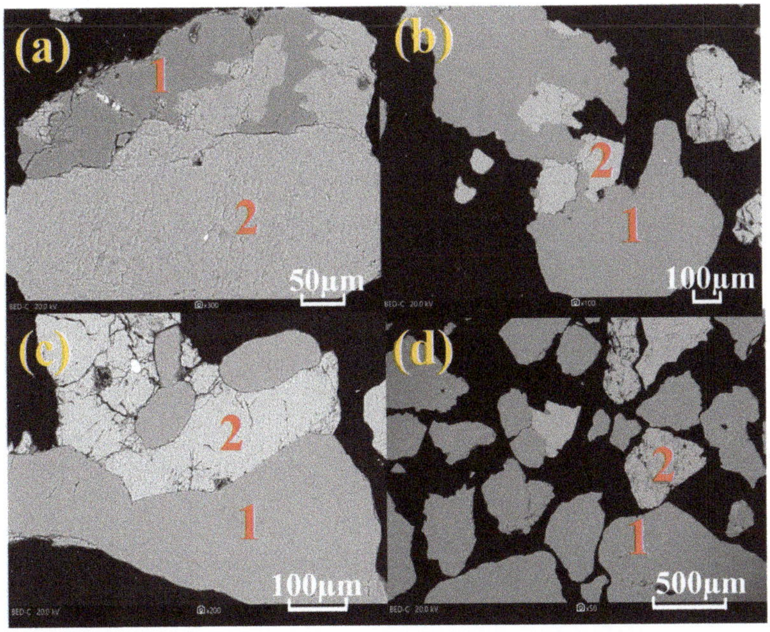

Figure 4. SEM pattern of raw ore. While area-1 is quartz and area-2 is Feldspar. ((**a**–**d**) are SEM images at different magnifications, electronic signal: backscattered electrons).

The optical microscope images of the pre-treated quartz are presented in Figure 5, revealing a significant presence of gas–liquid inclusions with particle sizes ranging from 1–10 μm. The majority of these inclusions exhibit a clustered spherical distribution, while a smaller portion is dispersed in spindle and bar shapes. Additionally, there are mineral inclusions present, which appear rounded and fusiform with particle sizes around 5–10 μm. Figure 6 displays the Raman microscope images and representative spectra of inclusions. Laser Raman analysis indicates characteristic Si-O vibration peaks at 465 cm^{-1}, O_2 peaks at 1567 cm^{-1}, and liquid -OH peaks at 3203 cm^{-1}, suggesting that the main components of the inclusions are O_2 and H_2O. Furthermore, Raman spectra also reveal traces of hydrocarbon fluid with characteristic peaks at 2970 cm^{-1}.

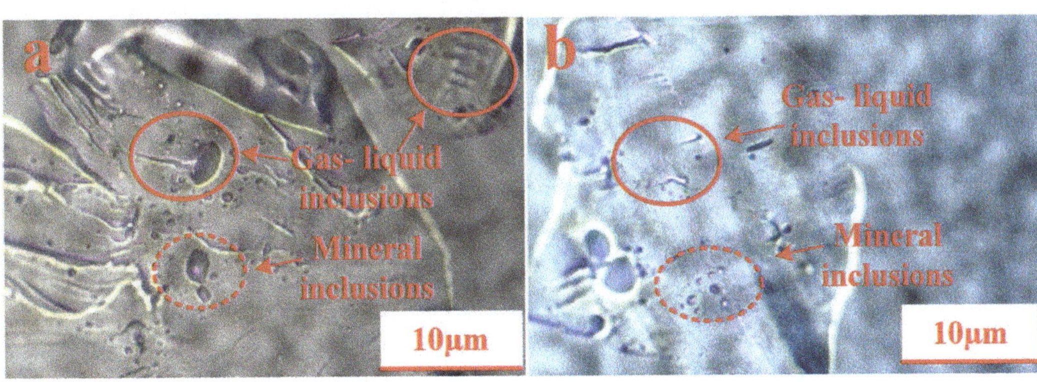

Figure 5. Optical microscope images of pre-treated quartz showing gas–liquid inclusions. ((**a**,**b**) are the images of inclusions observed at different locations).

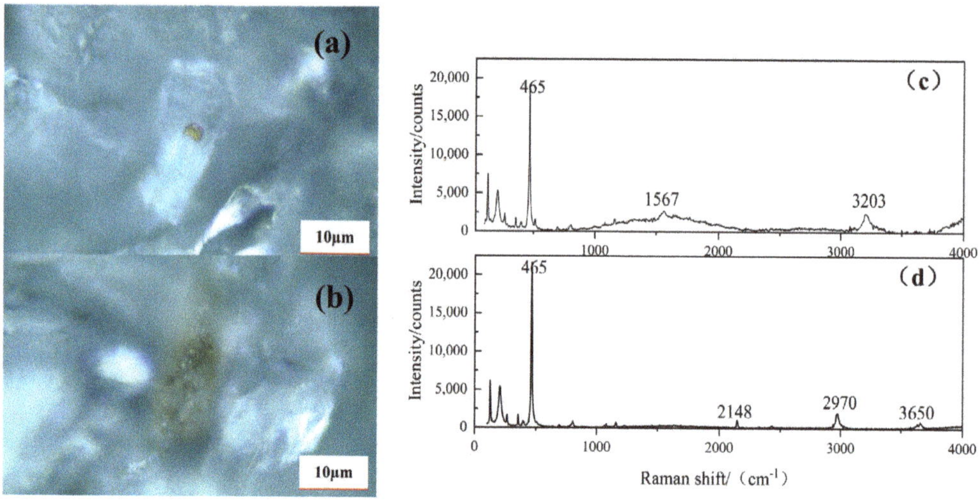

Figure 6. Raman image and spectrum. (The Raman spectral peak corresponding to Figure (**a**) is Figure (**c**), and the Raman spectral peak corresponding to Figure (**b**) is Figure (**d**)).

3.2. *Effect of Magnetic Separation Condition on Fe_2O_3 Removal*

The Fe_2O_3 content in the raw ore is 0.17%, which seriously affects the application of quartz, and magnetic separation is the most effective and environmentally friendly means of iron reduction. The purpose of the experiment of magnetic separation pulp concentration and magnetic field strength condition is to obtain the best magnetic separation process parameters. The results of SiO_2 recovery and Fe_2O_3 content of non-magnetic products are shown in Figure 7.

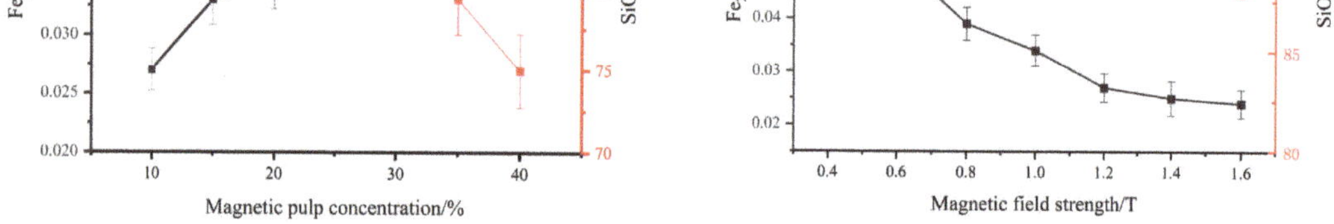

Figure 7. Effect of magnetic separation conditions on quartz products. ((**a**) is the magnetic field concentration curve, and (**b**) is the magnetic field intensity curve).

The pulp concentration test was set by controlling the magnetic concentration of 10%, 15%, 20%, 25%, 30%, 35% and 40% respectively, and the results of magnetic separation (Figure 7a) indicate that the recovery rate of quartz decreases with the increase of pulp concentration, and the rate of decline increases gradually. In addition, the content of Fe_2O_3 increases with the increase of pulp concentration. Notably, the Fe_2O_3 content reaches its minimum at a pulp concentration of 10%. This phenomenon can be attributed to higher slurry concentrations facilitating the adhesion of magnetic substances to non-magnetic surfaces, thereby elevating Fe_2O_3 content. Additionally, increased slurry density leads to quartz being more readily trapped among magnetic media, which in turn reduces SiO_2 recovery rates [22].

The magnetic field strength was controlled as 0.4 T, 0.6 T, 0.8 T, 1.0 T, 1.2 T, 1.4 T and 1.6 T, respectively, to set up the magnetic field strength test, and the analysis of magnetic field strength (Figure 7b) indicates that as the magnetic field intensity increases, both the concentration of Fe_2O_3 in non-magnetic products and the recovery rate of SiO_2 tend to decline. Moreover, when the magnetic field strength is greater than 1.2 T, the decline of Fe_2O_3 content in magnetic products tends to be gentle, but the recovery rate of quartz drops faster. This phenomenon may be attributed to the tendency for magnetic substances to form agglomerates under high magnetic field strengths, which leads to non-magnetic quartz minerals being inadvertently incorporated into magnetic products, thereby resulting in a loss of SiO_2 [23]. Furthermore, these results suggest that further increases in magnetic field intensity do not markedly enhance the reduction of iron impurities; thus, it is likely that residual iron impurities are embedded within the quartz crystal lattice.

Taking into account factors such as iron impurity content and quartz recovery rates, it is advisable to maintain the magnetic concentration at 10%–15% and control the magnetic field intensity at 1.2 T. Under optimal experimental conditions, the yield of non-magnetic products reaches 94.31%, while the iron oxide removal rate, as calculated by Equation (3), is determined to be 87.24%, indicating a significant efficacy in iron removal.

3.3. Effect of Microwave Treatment on Impurity Removal

NH_4Cl, NaCl, KCl and Na_2SO_4 were used as calcined dopants with the dosage of 2 wt%. After leaching, the conditional experimental results are shown in Table 3. It can be seen that after calcination with dopants, the impurity removal rate is obviously better than that of blank calcination, and the impurity residue is least when NH_4Cl is used as dopant. NaCl and Na_2SO_4 as dopants will cause a large amount of Na impurity residues, and KCl as a roasting dopant will cause a large amount of K^+ residues, which may be because the quartz crystal lattice changes during the roasting process, and because the dopant is in a molten state at high temperature, and the following reactions occur [16]:

$$KCl(s) \rightarrow K^+ + Cl^- \tag{4}$$

$$NaCl(s) \rightarrow Na^+ + Cl^- \tag{5}$$

$$Cl^- + 3a^{3+} \rightarrow aCl_3(g) \qquad (a = Al \text{ or } Fe) \tag{6}$$

$$2b^+ + O^{2-} \rightarrow b_2O \qquad (b = K \text{ or } Na) \tag{7}$$

$$b_2O + SiO_2 \rightarrow b_2SiO_3 \tag{8}$$

Table 3. The effect of dopants on the removal of impurities.

	Al/ppm	K/ppm	Fe/ppm	Na/ppm	Total/ppm
Undo	164.32	102.95	24.84	101.29	544.26
NH_4Cl	79.24	59.22	0.52	59.10	220.44
NaCl	70.42	11.01	14.84	145.82	303.41
KCl	85.45	221.11	18.19	4.72	395.40
Na_2SO_4	129.02	87.45	21.84	236.5	524.26

The molten Cl^- undergoes reactions with Al, Fe, and other elements within the quartz crystal lattice to generate ascending chloride gas. The specific reaction Equations (4)–(8) are presented to disrupt the crystal lattice and eliminate certain impurity elements. However, throughout the reaction process, the K^+ and Na^+ from the molten salt may react with O in the air or quartz, potentially infiltrating the lattice to form new silicate minerals. This can actually make it more difficult to remove impurities through acid leaching. The use of NH_4Cl as a dopant will decompose at high temperatures, leading to subsequent reactions.

$$NH_4Cl(s) \rightarrow NH_3(g) + HCl(g) \quad (9)$$

Compared with molten salts, NH_4Cl only decomposes into gas at high temperature, and it is not easy to introduce new impurities while destroying the quartz lattice, which can be removed by washing after calcination. In summary, NH_4Cl-assisted calcination with the best effect should be used in subsequent experiments.

According to the above experimental results, NH_4Cl (2 wt%) doping was used for microwave calcination to explore the effect of calcination time on the removal of main impurities. The calcination temperature was 900 °C, and the holding time was controlled for 0.5 h, 1.0 h, 1.5 h, 2.0 h and 2.5 h, respectively. The results after leaching were shown in Figure 8. It can be seen from the impurity removal rate curve that with the increase of holding time, the removal rate of all impurities increases, and tends to be flat after 2 h. The results show that increasing the holding time can indeed accelerate the leaching effect of Al, K, Fe and other elements, so the subsequent experiment should be carried out on the basis of constant temperature calcination for 2 h.

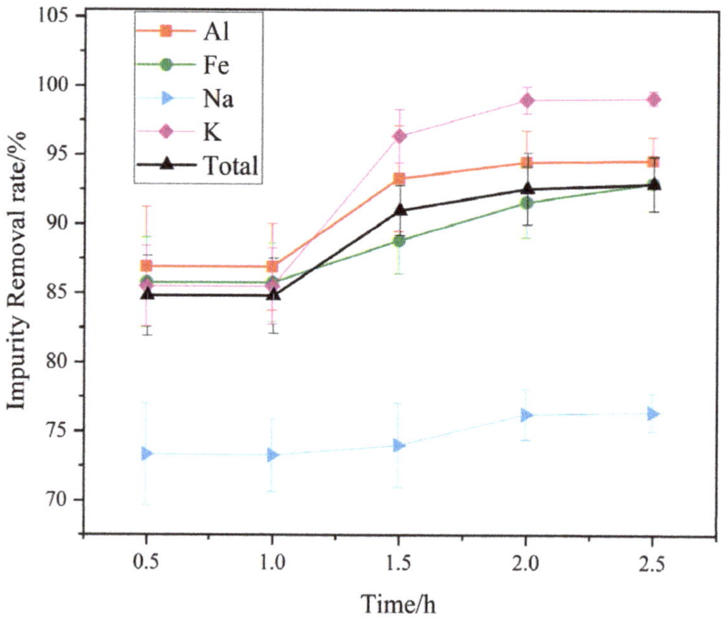

Figure 8. Effect of holding time on removal of impurities.

In order to explore the influence of calcination temperature on the removal effect of main impurities, the calcination temperature was controlled as 500 °C, 700 °C, 900 °C, 1100 °C, and the calcination time was 2 h. The experimental results of the calcination temperature are shown in Figure 9. It can be seen from the impurity removal rate curve that the removal rates of Al and K impurities gradually increase with the increase of temperature, and the removal effect is best at 1100 °C. The removal rates of Na and Fe showed a slightly fluctuating trend with the increase of temperature. Fe had the best removal effect at 900 °C, while Na had the best removal effect at 1100 °C. Considering the removal effect of all impurities, 1100 °C should be selected as the best calcination temperature.

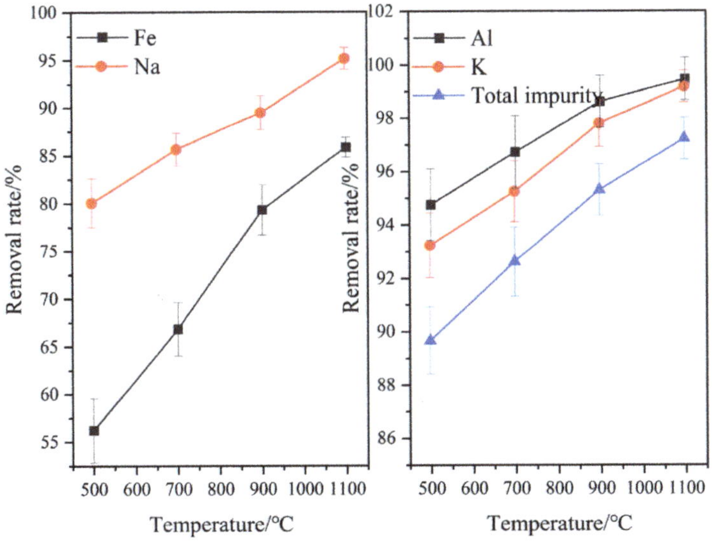

Figure 9. Effect of calcination temperature on removal of impurities.

In addition, we explored the effect of heating rate on the removal effect of main impurities, the heating rate was controlled at 25 °C/min and 50 °C/min respectively. The analysis of the heating rate results was shown in Table 4, which shows that the heating rate can increase the removal effect of total impurities to a certain extent, but the effect is not obvious compared with other experimental conditions. The temperature rise rate can be controlled to 50 °C/min.

Table 4. Effect of heating rate on removal of impurities.

	Al/ppm	K/ppm	Fe/ppm	Na/ppm	Total/ppm
25 °C/min	94.01	17.66	36.55	35.45	261.70
50 °C/min	83.42	12.92	26.62	38.56	219.28

Taking the above experimental results of calcination conditions into consideration, 2 wt% NH_4Cl was used as the dopant for 2 h by microwave calcination, the calcination temperature was 1100 °C, and the heating rate was controlled at 50 °C/min. Finally, high-purity quartz products with impurity content less than 100 ppm were prepared. ICP-OES test results are shown in Table 5.

Table 5. The contents of impurity elements of quartz after calcination by ICP-OES analysis.

Elements	Al	Ca	Cr	Cu	Fe	K	Li	Na	Others	Total
Content (ppm)	20.22	10.36	4.04	2.02	19.64	17.64	0.18	14.50	10.51	99.10

3.4. Surface Topography Analysis

The SEM in Figure 10 shows the surface of the quartz product. Prior to calcination, the quartz particles appear smooth without obvious cracks and holes (Figure 10a,b), while after calcination, clear cracks and corrosion pits can be observed on the surface of the quartz particles (Figure 10c,d). This is attributed to the high permeability of microwave radiation. During heating, powder materials are heated through conversion of micro electromagnetic waves into heat energy. Due to its good permeability to microwave radiation, gas-phase and liquid-phase inclusions within quartz crystal lattice exhibit strong wave absorption, resulting in selective heating by microwave radiation [24,25]. The rapid heating causes inclusions inside quartz to burst, leading to the formation of cracks on its surface which facilitate subsequent leaching and impurity removal. Simultaneously, as NH_4Cl undergoes thermal decomposition, the resulting HCl gas will react with impurities on the surface of quartz, leading to the formation of corrosion pit marks [26].

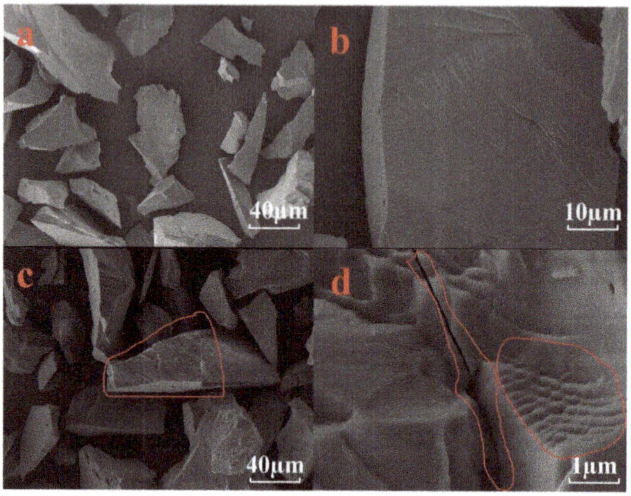

Figure 10. SEM pattern of quartz: (**a**,**b**) Quartz sample before calcination; (**c**,**d**) Quartz sample after calcination. (Electronic signal: secondary electrons).

3.5. *In-Situ XRD Analysis*

The in-situ XRD image of calcined quartz is shown in Figure 11a, and partial enlarged images are shown in Figure 10b,c. It can be observed from the figure that at 400 °C, the sample closely matches the standard diffraction peak of α-quartz (JCPDS: 01-078-4811). As temperature increases, the position of the main diffraction peak shifts within the range from 500 °C to 600 °C (as shown in Figure 11b,c), indicating a crystal transformation in progress. This aligns with recorded β-quartz (high-temperature quartz, JCPDS: 04-006-5049) transition temperatures in the existing literature [27,28], and this deviation gradually increases as calcination temperature rises. Wei et al.'s study on phase using density functional theory has demonstrated that such crystal phase transition facilitates interstitial impurity migration [29]. Upon original α-quartz overcoming energy barriers to transform into β-quartz, Si-O bond length within the lattice shortens, resulting in a larger change in interstitial volume caused by phase transition—thus promoting interstitial impurity migration to the surface and facilitating impurity removal.

Table 6 illustrates the variation of cell parameters throughout the heating process. It is evident that the (101) crystal surface of quartz initially decreases and subsequently increases as the temperature rises from 400 °C to 1100 °C, with the most significant change occurring between 500 °C and 600 °C, likely associated with the transformation from α-quartz to β-quartz. Furthermore, numerous studies have indicated that the peak shape in X-ray diffraction (XRD) is correlated with the morphology, volume, and particle size distribution of the unit cell. Additionally, a broader diffraction peak width corresponds to a smaller average unit cell size [30,31]. As shown in Table 6, there is a continuous reduction in half-peak width (FWHM) for quartz while both A-axis and C-axis dimensions are elongated; this suggests that as temperature increases, quartz cells undergo stretching, leading to an increase in average size. Concurrently, there is a gradual decrease in c/a ratio indicating that elongation along the axis occurs at a slower rate compared to that along the c axis. These observations further substantiate that during calcination processes, alterations occur within quartz's crystal cell parameters which may facilitate impurity incorporation into lattice defects [32].

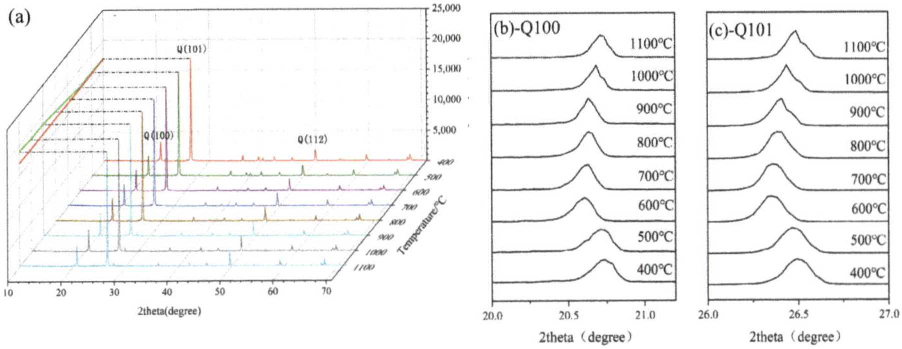

Figure 11. In-situ XRD image of quartz product ((**b**,**c**) are partial images of (**a**)).

Table 6. Cell parameters at different temperatures during calcination.

Temperature/°C	2θ-(101)/°	FWHM	a/nm	c/nm	c/a
400	26.449	0.162	4.951	5.431	1.0970
500	26.408	0.154	4.966	5.441	1.0957
600	26.259	0.146	4.998	5.463	1.0930
700	26.231	0.134	5.002	5.462	1.0920
800	26.220	0.129	5.000	5.460	1.0920
900	26.241	0.115	5.000	5.458	1.0916
1000	26.248	0.111	5.000	5.457	1.0914
1100	26.250	0.109	5.000	5.456	1.0912

In addition, the analysis in Figure 11 also shows that the characteristic peak intensity of Q (101) crystallographic plane increases slowly with the increase of temperature. This indicates that the main crystal plane diffraction intensity of quartz is increasing. At the same time, there are double distinct characteristic diffraction peaks of aluminum silicate (JCPDS: 01-078-0436) at 73.206° and 75.477° compared to quartz, as shown in Figure 12. It can be observed that the area of the characteristic diffraction peaks of aluminum silicate decreases with increasing calcination temperature. This indicates that the crystalline phase of aluminum silicate also undergoes changes during the calcination process, which may be beneficial for subsequent impurity removal.

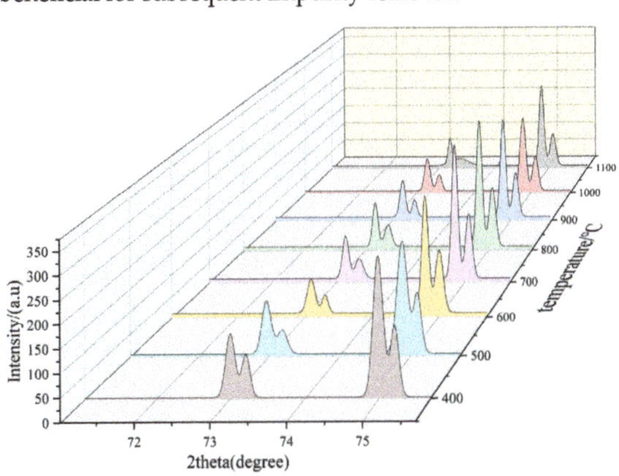

Figure 12. Locally enlarged image of XRD pattern (72°–75.7°).

3.6. Thermogravimetric Mass Spectrometry Analysis

The TG curve of quartz products is depicted in Figure 13, demonstrating the excellent thermal stability and consistent weight loss pattern of quartz. The loss on ignition from 30 °C to 1100 °C is only 0.21% (Figure 13a), indicating that the loss is likely attributed to bonded water within the quartz. In contrast, Figure 13b illustrates the TG curve of calcination with NH_4Cl added as a dopant. Mass spectrometry analysis (Figure 14) reveals a peak weight loss rate at 251.9 °C. Prior to this temperature, sample weight loss primarily results from NH_4Cl decomposition, yielding NH_3 and HCl. Subsequently, under the influence of HCl and high temperatures, impurities within the sample are corroded and inclusions are disrupted, leading to sequential production of H_2O, CO, and alkane gases. Additionally, due to its low decomposition temperature, resulting HCl can react with impurities such as aluminum silicate and compounds of iron or titanium on surface or within cracks (Equations (9) and (10)), causing vaporization and escape of these impurities; however, their low concentration makes them difficult to detect during mass spectrometry analysis.

$$6HCl + a_2O_3 \rightarrow 2aCl_3(g) + 3H_2O(g) \qquad (a = Al \text{ or } Fe) \qquad (10)$$

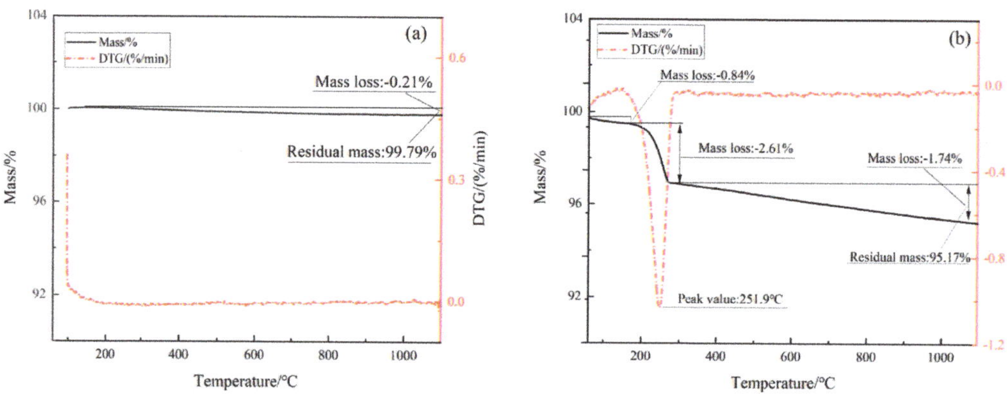

Figure 13. The TG-DTG curve of the sample: (**a**) Curves of individually calcined quartz; (**b**) Curve of mixed calcination of quartz and NH4Cl (2 wt%).

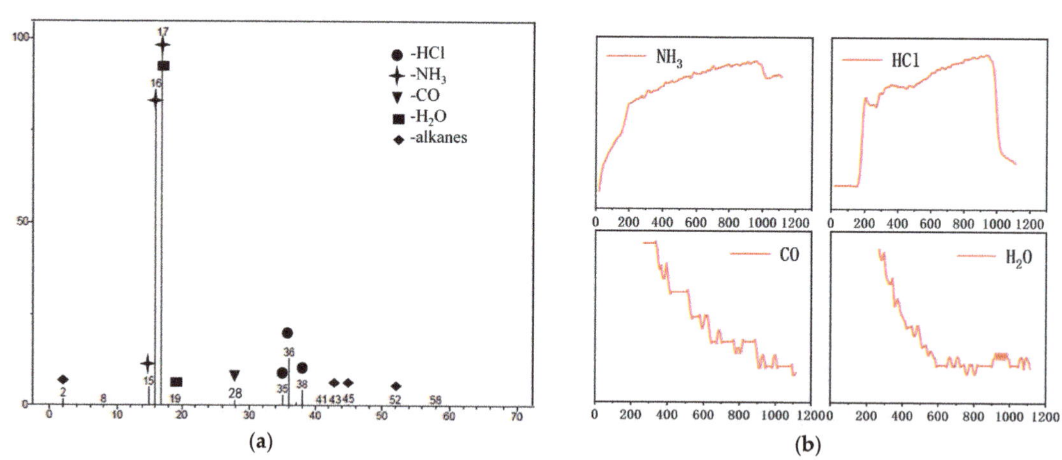

Figure 14. MS curve of mixed calcination of quartz and NH4Cl (2 wt%): (**a**) Mass spectrometric analysis of gas produced during calcination; (**b**) The gas content curve varies over time.

Compared with the quartz calcined alone, the mass loss of the calcined sample is 2.83%, which is much higher than that of ordinary calcined sample. This proves that the calcined sample doped with NH_4Cl can significantly increase the fracture effect of inclusions in quartz.

3.7. Model and Discussion of Calcination Process

According to the analysis results of SEM, in-situ XRD, and TG-MS, the following model of the calcination process can be verified (see Figure 15). The TG-MS curve of calcination indicates that NH4Cl decomposes into NH3 and HCl gases at lower microwave heating temperatures. As the temperature increases, the crystal cell parameters of quartz expand, leading to an enlargement of the crystal cell volume and migration of impurities within lattice defects, as evidenced by in-situ XRD results. Subsequently, HCl reacts with surface impurities on quartz such as Al and Fe, disrupting the crystal lattice and resulting in the formation of water and chloride compounds. Concurrently, in-situ XRD analysis reveals a reduction in the diffraction peak area corresponding to aqueous aluminum silicate within quartz, signifying alterations in its crystal lattice structure. It is plausible that larger inclusions within quartz are selectively heated due to their superior wave absorption properties, causing fracture and localized fusion of these inclusions. According to the existing literature, aluminum chloride and ferric chloride have boiling points of 178 °C and 316 °C, respectively; thus, when temperatures reach 200–300 °C [33], chlorides sublime while removing certain impurities from the quartz surface—evident through cracks and corrosion pits observed in SEM images. With further temperature elevation, gas–liquid inclusions escape through these fractures while mineral inclusions (predominantly aluminosilicates) continue reacting with HCl present in the fissures before sublimating at elevated temperatures to facilitate comprehensive removal of internal impurities.

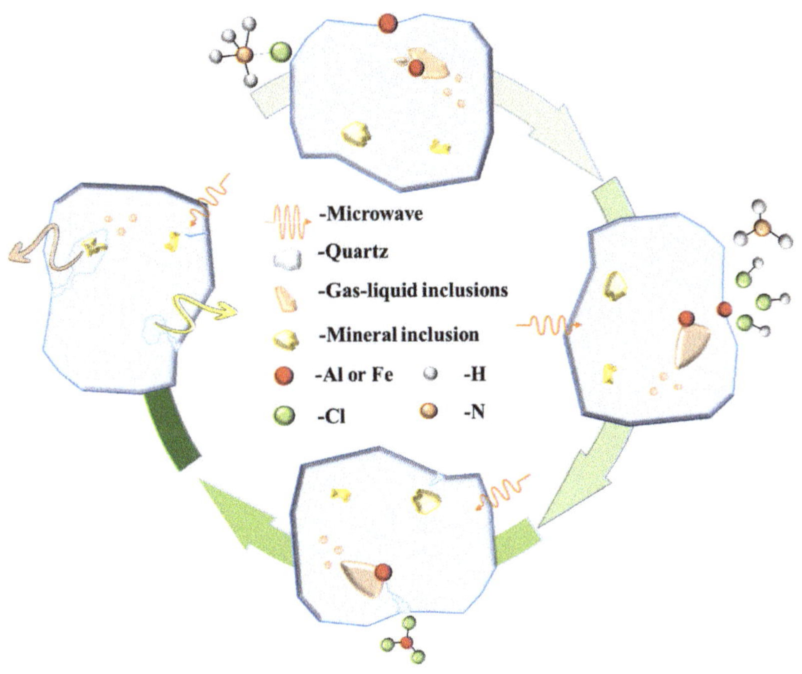

Figure 15. Model of microwave calcination process.

4. Conclusions

In this study, high-purity quartz with an impurity content of less than 100 ppm was successfully obtained from Yueyang river sand through a series of processes including magnetic separation pretreatment, flotation, roasting, and leaching. The factors influencing the efficacy of magnetic separation and roasting were thoroughly examined, and the mechanism of the calcination process was further investigated. Specific conclusions are as follows:

1. The pulp concentration of the magnetic separation test should be controlled at 15% and the magnetic field strength is 1.2 T, based on which the iron removal rate can be increased to 87.24%.
2. The optimal calcination conditions for the raw ore were as follows: a doping agent of 2 wt% NH_4Cl was used, microwave calcination was conducted for 2 h at a temperature of 1100 °C, and the heating rate was controlled at 50 °C/min. By means of chemical analysis, laser Raman analysis, and SEM analysis, the characteristics of inclusions in the original ore were determined. The main components of these inclusions are H_2O, O_2, CO_2, and hydrocarbon fluid impurities.
3. Scanning electron microscope (SEM) photos taken before and after calcination, along with X-Ray diffraction (XRD) and thermogravimetric mass spectrometry (TG-MS) analysis conducted during the calcination process, indicate that under the influence of high-temperature phase transition and NH_4Cl, the inclusions inside quartz undergo rupture, resulting in cracks and corrosion pits, and impurities are removed after sublimation.
4. Mass spectrometry shows that the gas produced after the rupture of the inclusion mainly consists of water, carbon oxides, and alkanes. This aligns with the analysis of the Raman spectrum and the formation conditions of river sand.

In summary, previous studies in this direction were mostly based on the analysis of quartz after roasting, and the mechanism of the roasting heating process was lacking. The innovation of this study is rooted in the effective utilization of waste resources through the extraction of high-purity quartz from waste river sand. The types of impurities present in river sand were characterized using spectroscopy and microscopy techniques, thereby validating the feasibility of purifying high-purity quartz from this source. Furthermore, for the first time, in situ X-ray diffraction (XRD) and thermogravimetric-mass spectrometry (TG-MS) were employed to investigate the heating process of quartz inclusion removal and a model of inclusions removal during calcination was established, offering a theoretical framework for the utilization of similar resource types.

Author Contributions: Conceptualization, methodology, resources, writing—review and editing, H.H.; investigation, data curation, writing—original draft preparation, N.Z. All authors have read and agreed to the published version of the manuscript.

Funding: This research received no external funding.

Data Availability Statement: Data are contained within the article.

Acknowledgments: We are grateful to the Engineering Research Center of the Ministry of Education for Carbon Emission Reduction in Metal Resource Exploitation and Utilization of Central South University for technical support.

Conflicts of Interest: The authors declare no conflicts of interest.

References

1. Long, H.; Zhu, D.; Pan, J.; Li, S.; Yang, C.; Guo, Z. Advanced Processing Techniques and Impurity Management for High-Purit Quartz in Diverse Industrial Applications. *Minerals* **2024**, *14*, 571. [CrossRef]

1. Pan, X.D.; Li, S.Q.; Li, Y.K.; Guo, P.H.; Zhao, X.; Cai, Y.S. Resource, characteristic, purification and application of quartz: A review. *Miner. Eng.* **2022**, *183*, 107600. [CrossRef]
2. Lin, M.; Liu, Z.; Wei, Y.; Liu, B.; Meng, Y.; Qiu, H.; Lei, S.; Zhang, X.; Li, Y. A Critical Review on the Mineralogy and Processing for High-Grade Quartz. *Min. Metall. Explor.* **2020**, *37*, 1627–1639. [CrossRef]
3. Wang, L.; Wang, G.D.; Ge, P.; Sun, W.; Tang, H.H.; Hu, W.J.H. Activation mechanisms of quartz flotation with calcium ions and cationic/anionic mixed collectors under alkalescent conditions. *Colloid Surf. A* **2022**, *632*, 127771. [CrossRef]
4. Li, Y.K.; Li, S.Q.; Pan, X.D.; Zhao, X.; Guo, P.H. Pre-concentration of quartz from sea sand through superconducting high gradient magnetic separation technology. *Sep. Sci. Technol.* **2023**, *58*, 822–834. [CrossRef]
5. Wang, L.; Liu, R.Q.; Hu, Y.H.; Liu, J.P.; Sun, W. Adsorption behavior of mixed cationic/anionic surfactants and their depression mechanism on the flotation of quartz. *Powder Technol.* **2016**, *302*, 15–20. [CrossRef]
6. Liu, C.F.; Wang, W.T.; Wang, H.; Zhu, C.Y.; Ren, B. A Review on Removal of Iron Impurities from Quartz Mineral. *Minerals* **2023**, *13*, 1128. [CrossRef]
7. Sekulic, Z.; Bartulovic, Z.; Mihajlovic, S.; Ignjatovic, M.; Savic, L.; Jovanovic, V.; Nisic, D. The choice of high gradient magnetic separation processes for removal of Fe_2O_3 carriers from quartz raw material. *Gospod. Surowcami Miner. Miner. Resour. Manag.* **2017**, *33*, 93–106. [CrossRef]
8. Müller, A.; Koch-Müller, M. Hydrogen speciation and trace element contents of igneous, hydrothermal and metamorphic quartz from Norway. *Mineral. Mag.* **2009**, *73*, 569–583. [CrossRef]
9. Dong, X.L.; Xiong, Y.X.; Wang, N.; Song, Z.; Yang, J.; Qiu, X.M.; Zhu, L.H. Determination of trace elements in high-purity quartz samples by ICP-OES and ICP-MS: A normal-pressure digestion pretreatment method for eliminating unfavorable substrate Si. *Anal. Chim. Acta* **2020**, *1110*, 11–18. [CrossRef] [PubMed]
11. Götze, J.; Pan, Y.M.; Müller, A. Mineralogy and mineral chemistry of quartz: A review. *Mineral. Mag.* **2021**, *85*, 639–664. [CrossRef]
12. Gawel, B.A.; Ulvensoen, A.; Lukaszuk, K.; Muggerud, A.M.F.; Erbe, A. In situ high temperature spectroscopic study of liquid inclusions and hydroxyl groups in high purity natural quartz. *Miner. Eng.* **2021**, *174*, 107238. [CrossRef]
13. Gulbin, Y.L.; Abdrakhmanov, I.A.; Gembitskaya, I.M.; Vasiliev, E.A. Oriented Microinclusions of Al-Fe-Mg-Ti Oxides in Quartz from Metapelitic Granulites of the Bunger Hills, East Antarctica. *Geol. Ore Depos.* **2023**, *65*, 656–668. [CrossRef]
14. Zhong, T.; Yu, W.; Shen, C.; Wu, X. Research on Preparation and Characterisation of High-purity Silica Sands by Purification of Quartz Vein Ore from Dabie Mountain. *Silicon* **2022**, *14*, 4723–4729. [CrossRef]
15. Li, Y.; Ma, Q.; Xia, Z.; Li, W.; Lei, S. Influences of Na_2CO_3 roasting and H_3PO_4 hot-pressure leaching on the purification of vein quartz to achieve high-purity quartz. *Hydrometallurgy* **2023**, *218*, 106065. [CrossRef]
16. Lin, M.; Pei, Z.; Li, Y.; Liu, Y.; Wei, Z.; Lei, S. Separation mechanism of lattice-bound trace elements from quartz by KCl-doping calcination and pressure leaching. *Miner. Eng.* **2018**, *125*, 42–49. [CrossRef]
17. Shao, H.; Zang, F.; Ji, M.; Yu, M. Prepare and Mechanism of High Purity Quartz by Alkali Corrosion and Acid Leaching Process Using Vein Quartz. *Silicon* **2022**, *14*, 12475–12483. [CrossRef]
18. Guo, S.J.; Ai, G.T.; Zhao, J.J.; Yang, S.C.; Wei, K.X.; Ma, W.H. Removal of hydroxyl impurities in vein quartz by vacuum roasting. *Vacuum* **2024**, *222*, 113049. [CrossRef]
19. Li, F.F.; Jiang, X.S.; Zuo, Q.X.; Li, J.W.; Ban, B.Y.; Chen, J. Purification Mechanism of Quartz Sand by Combination of Microwave Heating and Ultrasound Assisted Acid Leaching Treatment. *Silicon* **2021**, *13*, 531–541. [CrossRef]
20. Oghbaei, M.; Mirzaee, O. Microwave versus conventional sintering: A review of fundamentals, advantages and applications. *J. Alloys Compd.* **2010**, *494*, 175–189. [CrossRef]
21. Yuan, B.; Li, C.; Liang, B.; Lü, L.; Yue, H.; Sheng, H.; Ye, L.; Xie, H. Extraction of potassium from K-feldspar via the $CaCl_2$ calcination route. *Chin. J. Chem. Eng.* **2015**, *23*, 1557–1564. [CrossRef]
22. Li, Y.; Li, S.; Zhao, X.; Pan, X.; Guo, P. Separation and purification of high-purity quartz from high-silicon iron ore tailing: An innovative strategy for comprehensive utilization of tailings resources. *Process Saf. Environ. Prot.* **2023**, *169*, 142–148. [CrossRef]
23. Zheng, X.; Cheng, W.; Cheng, J.; Shuwak, S.; Yue, X.; Li, B.; Wang, X.; Zhou, S.; Lu, D.; Wang, Y. Enhancing the separation of refractory weakly magnetic minerals with magnetic fluid in high gradient magnetic separation: Conversion of competing capture to mechanical entrainment. *Sep. Purif. Technol.* **2024**, *335*, 126273. [CrossRef]
24. Song, W.F.; Jiang, X.S.; Chen, C.; Ban, B.Y.; Wan, S.M.; Chen, J. Purification of Quartz Via Low-Temperature Microwave Chlorinated Calcination Combined with Acid Leaching and its Mechanism. *Silicon* **2023**, *15*, 971–981. [CrossRef]
25. Hou, Q.; Jing, L.; Yin, R.; Chen, L.; Lu, L.; Ji, F. Study on gas-liquid inclusions in quartz sand under microwave field. In Proceedings of the 2012 International Conference on Chemical Engineering, Metallurgical Engineering and Metallic Materials, CMMM 2012, Kunming, China, 12–13 October 2012; pp. 689–693.
26. Ren, Z.; Liu, Z.; Liu, Y.; Gao, H.; Wu, F.; Song, Y. The Effects of Calcination-Water Quenching on Quartz Purification and Its Mechanism. *Min. Metall. Explor.* **2023**, *40*, 2519–2527. [CrossRef]

27. Wright, A.F.; Lehmann, M.S. Structure of Quartz at 25 and 590 Degree C Determined by Neutron Diffraction. *J. Solid State Cher* **1981**, *36*, 371–380. [CrossRef]
28. Zhang, H.W.; Guo, S.J.; Wu, J.J.; Wu, D.D.; Wei, K.X.; Ma, W.H. Effect of quartz crystal structure transformations on the remova of iron impurities. *Hydrometallurgy* **2021**, *204*, 105715. [CrossRef]
29. Wei, Z.; Li, Y.; Li, P.; Pan, L.; Li, W.; Hu, X.; Gu, Y. New insight into the plastic deformation mechanisms during the SiO_2 phas transition process. *RSC Adv.* **2024**, *14*, 6262–6269. [CrossRef]
30. VanLeeuwen, B.K.; Darling, K.A.; Koch, C.C.; Scattergood, R.O.; Butler, B.G. Thermal stability of nanocrystalline Pd81Zr19. *Act Mater.* **2010**, *58*, 4292–4297. [CrossRef]
31. Sánchez-Bajo, F.; Ortiz, A.L.; Cumbrera, F.L. Novel analytical model for the determination of grain size distributions in nanocry talline materials with low lattice microstrains by X-Ray diffractometry. *Acta Mater.* **2006**, *54*, 1–10. [CrossRef]
32. Karunadasa, K.S.P.; Manoratne, C.H.; Pitawala, H.M.T.G.A.; Rajapakse, R.M.G. The composition, unit cell parameters an microstructure of quartz during phase transformation from α to β as examined by in-situ high-temperature X-Ray powde diffraction. *J. Phys. Chem. Solids* **2018**, *117*, 131–138. [CrossRef]
33. Ihara, Y.; Ohgame, H.; Sakiyama, K.; Hashimoto, K. The corrosion behaviour of iron in hydrogen chloride gas and gas mixtures c hydrogen chloride and oxygen at high temperatures. *Corros. Sci.* **1981**, *21*, 805–817. [CrossRef]

minerals

MDPI

·ticle

·reconcentrating Ultrafine Ilmenite Tailings Using a .aboratory-Scale Reflux Classifier

ıenqiang Liu [1], Zhenhua Su [2], Bing Liu [2], Yuhua Wang [1], Yuxin Zhang [1], Xuqun Zhong [2], Kangkang Chen [3], .aoxing Hu [3] and Dongfang Lu [1,*]

1 School of Minerals Processing and Bioengineering, Central South University, Changsha 410083, China; 215601035@csu.edu.cn (Z.L.); wangyh@csu.edu.cn (Y.W.); 245603045@csu.edu.cn (Y.Z.)
2 Guangxi Zhongiin Lingnan Mining Co., Ltd., Laibing 545900, China; m19987529673@163.com (Z.S.); gikevin@163.com (B.L.); wytkcu@163.com (X.Z.)
3 BGRIMM Technology Group, Beijing 100160, China; chenkangkang@bgrimm.com (K.C.); huxiaoxing@bgrimm.com (X.H.)
* Correspondence: ludongfang@csu.edu.cn; Tel.: +86-153-9990-6602

Abstract: China is rich in reserves of titanium, but a large amount of titanium resources is lost in the ultrafine tailings, and it is challenging to treat the ilmenite contained in ultrafine ore. The reflux classifier (RC), a novel gravity concentration technology, has been applied in the preconcentration of ultrafine ilmenite in this study. During this process, the feasibility of using RC for preconcentration of ultrafine ilmenite was explored through theory and conditional experiments. After one-stage preconcentration using RC, the ultrafine ilmenite ore with a TiO_2 grade of 8.77% can be concentrated into a product with a TiO_2 grade of 20.3% and a recovery rate of 82.8%. The tailings grade is as low as 2.44%, and the yield reaches 62.6%. The separation efficiency achieves 50.0%. Experimental results demonstrate that utilizing RC for the preconcentration of ultrafine ilmenite can avoid the influence of weakly magnetic gangue and achieve better results compared to a magnetic separator. Therefore, RC offers a more effective and affordable method for preconcentrating ultrafine ilmenite ore.

Keywords: ultrafine ilmenite; physical separation; gravity concentration; preconcentrating

1. Introduction

China possesses vast reserves of titanium resources, comprising over 35% of the world's total reserves. The majority of these resources, over 90% [1], are primarily concentrated in the Panzhihua area of Sichuan province, predominantly in the form of ilmenite. Common gangue minerals include feldspar and augite, with a minor quantity of olivine present [2,3]. Due to the valuable minerals being embedded in the gangue in the form of extremely fine particles, raw materials must be finely ground to ensure product quality, resulting in the generation of a significant volume of fine particle materials [3–5]. Consequently, a substantial quantity of fine-grained ilmenite is lost in the tailings, resulting in the wastage of valuable titanium resources [6–8]. The commonly used techniques for ilmenite concentration are magnetic separation and flotation [5,9]. In the case of magnetic separation, to attain optimal beneficiation products when separating ultrafine ilmenite, it is imperative to utilize a superconducting magnetic separator capable of generating a maximum magnetic field intensity of up to 5 T [7]. Indeed, this poses a substantial challenge given that conventional magnetic separators encounter difficulties in effectively separating ultrafine ilmenite particles [10–12]. Additionally, if the content of ilmenite in the ultrafine

 https://doi.org/10.3390/min14111125

raw material is low, the presence of fine-grained gangue minerals, especially augite and chlorite, can exert a considerable adverse impact on both magnetic separation [13] and flotation processes [14–17].

Investigating the viability of utilizing gravity concentration techniques for the beneficiation of ultrafine ilmenite represents a valuable pursuit. However, it is crucial to acknowledge that fine-grained minerals in a liquid medium tend to increase slurry viscosity and are highly susceptible to variations in settling velocity caused by particle size. The gravity separation of ultrafine particles poses a significant challenge for equipment.

Professor Galvin [18] from Newcastle University has introduced an innovative gravity separation device known as the reflux classifier (RC). This separation apparatus has demonstrated exceptional performance in the field of coal beneficiation [19–22] and is also gaining recognition in the domain of iron ore beneficiation. Amariei [23] employed iron tailings, which had been processed through a spiral chute, as the feed material for a pilot-scale RC. The sample exhibited a particle size distribution of less than 150 μm, and the spaces of the inclined channels are 6 mm. Ultimately, the recovery rate of total iron in the resulting concentrate product exceeded 80%. However, the recovery rate for fine particles (−40 μm) within the concentrate was observed to be below 40%. D. M. Hunter [24] conducted experiments using a laboratory-scale RC ($RC^{TM}100$ with 6 mm spacing inclined channels) for hematite and siliceous gangue separation at ultrafine sizes. The initial feed contained 31.9% of particles with a size of −20 μm and exhibited an Fe_T grade of 32.0%. The concentrate product achieved an 80.2% recovery rate with an Fe_T grade of 66.1%. However, it is worth noting that the recovery rate for the −20 μm fraction was only 57.0%, significantly lower than that of other sizes. Rodrigues [25] conducted tests using the laboratory-scale RC for the beneficiation of two itabirite iron ore mines in Brazil. The ore samples had a particle size distribution of less than 150 μm, with approximately 20% of particles in the −20 μm range. However, the recovery rates for both materials in the −20 μm particle size range were below 35%.

Based on the aforementioned reports, it is apparent that the mentioned RC did not deliver satisfactory results for the ultrafine grain in iron ore. To enhance the separation efficiency for ultrafine particles, an RC with an inclined channel spacing of 1.5 mm was tested for the preconcentration of ultrafine ilmenite. Additionally, modifications were made to the structural proportions of the RC. As previously mentioned, the ratio of the length of inclined channels to the height of the fluidized bed column in the $RC^{TM}100$ was 1:1. In this study, the height of the fluidized bed was reduced, resulting in an increased ratio of the length of inclined channels to the height of the column, reaching 4.5:1. This adjustment emphasizes the pivotal role played by the inclined channels in the beneficiation process.

2. Theory

2.1. Hydrodynamic Model in the Fluidized Bed

Different particles with varying sizes and densities have distinct settling velocities in the slurry due to the medium resistance effect. In fluidized beds, the settling terminal velocity equation can be used to establish a relationship between free settling velocity. The relationship between hindered settling terminal velocity and free settling terminal velocity is as follows [26,27]:

$$u_0 = \frac{d^2(\rho_s - \rho)g}{18\mu + 0.61d[(\rho_s - \rho)\rho g d]^{1/2}}, \tag{1}$$

$$u_t = u_0(1 - \lambda)^n, \tag{2}$$

where u_0 is the terminal velocity of free settling, u_t is the terminal velocity of hindered settling, d is the diameter of the particle, ρ_s is the density of the particle, ρ is the density of the fluid, μ is the viscosity of the fluid, g is the gravitational acceleration, and λ is the volume fraction of solids.

According to Equation (1), it is evident that particles with differing densities exhibit varying settling velocities, a crucial factor in fluidized beds for achieving particle separation based on density. In addition, a higher density of the fluidized bed, serving as the separation medium, proves advantageous for separating particles with different densities.

The solid volume fraction is proportional to the overall density of the liquid–solid fluidized bed; the higher the overall density of the liquid–solid fluidized bed, the higher the solid volume fraction. However, a higher overall density of the liquid–solid fluidized bed is more conducive to the separation of particles with different densities. According to Equation (2), the higher the solid volume fraction, the lower the terminal settling velocity of the particles, which will result in an extended sorting time. Hence, conducting experiments to strike a balance between the factors influencing bed concentration and density is imperative for exploring optimal separation conditions.

2.2. Hydrodynamic Model in the Inclined Channel

A planar diagram of the inclined channel is presented in Figure 1 [18], where the fluid passing through it exhibits laminar flow. The velocity distribution of the flow in the direction of thickness is parabolic. This assumption is made while considering the thickness of the channel (z).

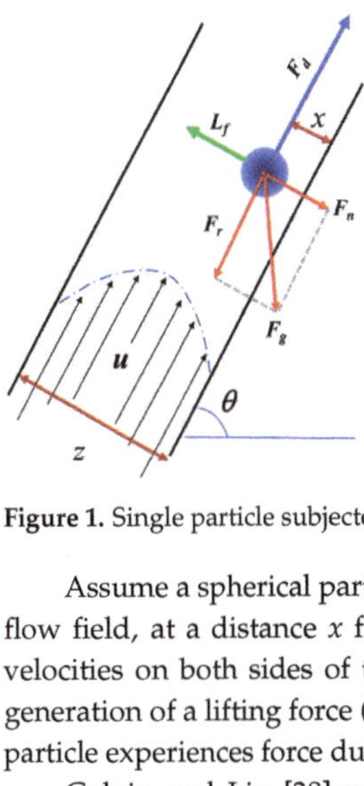

Figure 1. Single particle subjected to laminar flow and a high shear rate.

Assume a spherical particle of diameter d and density ρ_s is present in a non-uniform flow field, at a distance x from the upward-facing surface. In this scenario, the flow velocities on both sides of the particle differ in the normal direction, resulting in the generation of a lifting force (L_f) acting perpendicular to the direction of u. Therefore, the particle experiences force due to gravity, buoyancy, drag force (F_d), and lifting force (L_f).

Galvin and Liu [28] revised and integrated the studies conducted by King and Leighton [29], and Vance and Moulton [30]. Based on their work, they proposed the equation for lifting force (L_f) and subsequently verified it.

$$L_f = 0.0567\frac{\rho^{1.8}\gamma^{2.8}d^{5.6}}{\mu^{0.8}} = 0.0567Re_s^{0.8}\rho\gamma^2 d^4, \quad (3)$$

where Re_s is the shear Reynolds number, γ is the shear rate, and U' is the water velocity in the inclined channel, m/s.

$$Re_s = \rho\gamma d^2/\mu, \quad (4)$$

$$\gamma = \frac{6U'}{z}\left(1 - \frac{2x}{z}\right), \quad (5)$$

where z is the spacing of the inclined channel, and x is the distance from the particle center to the channel surface.

Hence, in the inclined channel, particles with densities exceeding the critical density ρ_s are prone to settle downwards, while particles with densities lower than ρ_s tend to be transported upward. If two particles of different densities are moving upwards within the inclined channel, lighter particles are more likely to be carried all the way to the overflow, while heavier particles are inclined to settle on the lower side of the inclined channel and return to the fluidized bed.

As illustrated in Figure 2, particles with different densities are ejected to varying distances. Yanfeng Li et al. [31] conducted computations and experimental verification of the particle motion distance along the channel (L_P) for densities at different apparent velocities.

$$L_P = \frac{z}{\cos\theta}\left(\frac{U'}{u_0} - 1\right), \tag{6}$$

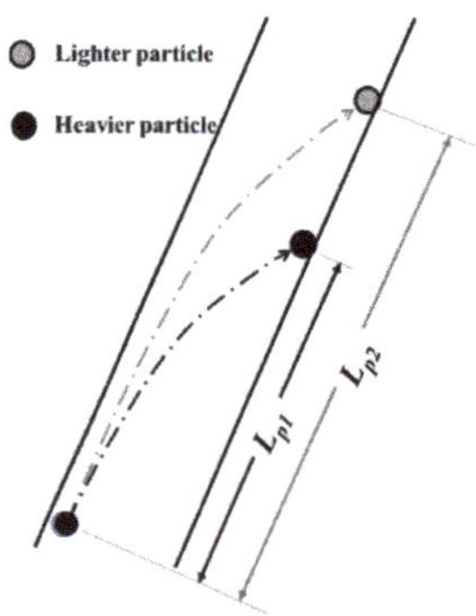

Figure 2. The trajectory of particles in the inclined channel.

Figure 3 illustrates the L_P values of particles of varying sizes and densities flowing through the inclined channel, and the curve in Figure 3 is calculated using Equation (6). In the inclined channel, as the particle diameter decreases, the difference in L_P values caused by the density becomes larger, resulting in a wider range of separating intervals for particles of different densities. Moreover, if the particles are extremely fine, high-density particles are more likely to settle near the entrance of the inclined channel, achieving separation from particles of different densities.

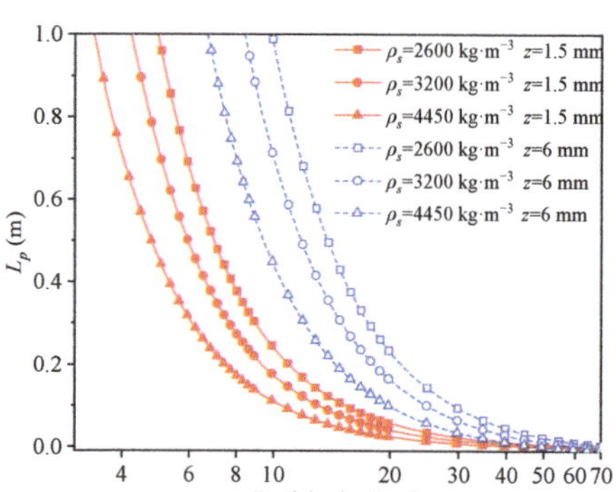

Figure 3. Particle motion distance along the channel.

It can be seen from Figure 3 that as the spacing of inclined channels increases, the value of L_p increases. With the channels' limited length, effectively capturing and recovering ultrafine heavy minerals becomes a challenging task. Separating ultrafine particles of varying densities becomes difficult, and the majority of fine particles end up being lost in the overflow. This factor contributes to the lower recovery rate of ultrafine particles in iron ore processing when wide-spaced inclined channels are utilized in the RC.

In conclusion, closely spaced inclined channels hold promise for separating ultrafine ilmenite effectively. Additionally, the apparent velocity within these channels will impact particle separation efficiency. Similarly, the concentration and density of the fluidized bed, as well as the velocity of the fluidization water, will significantly affect separation efficiency. Consequently, this study will focus on factors such as underflow rate, fluidization water velocity, and equipment processing capacity, all of which influence the parameters mentioned above.

3. Experiment Design

3.1. Equipment

The apparatus employed in this study is a laboratory-scale reflux classifier, as illustrated in Figure 4. The middle part is a prism and the inclined channel has an inclination angle of 70 degrees with respect to the horizontal plane. The inclined chamber is divided into 23 evenly distributed channels by 22 stainless steel plates, each with the same spacing and dimensions.

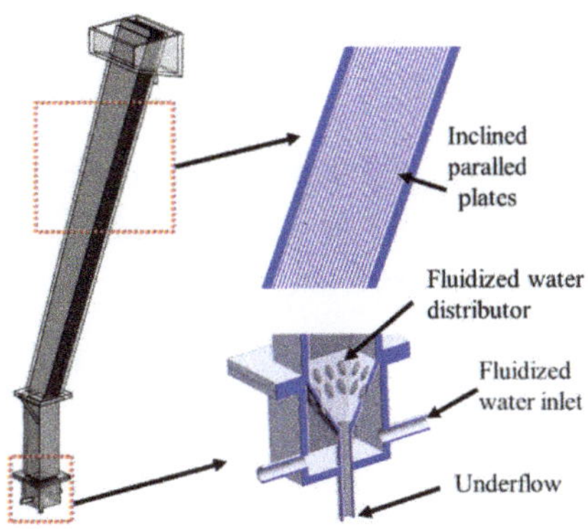

Figure 4. Laboratory-scale reflux classifier.

The equipment can be partitioned into three primary sections: the top inclined channels; the central vertical quadrangular prism, which serves as the fluidized bed separation zone; and the lower section, which contains the water distribution and underflow discharge port. The underflow is utilized to evacuate the heavy particles settled at the bottom of the fluidized bed, and its discharge capability is regulated by a peristaltic pump. Water from the tank is pressurized and sent via a pump into the distribution chamber, subsequently providing fluidized water uniformly into the fluidized bed via vertical minute holes present on the distribution plate. The particles present on the upper layer of the fluidized bed then progress towards the inclined channel for further separation, ultimately resulting in low-density product overflowing from the uppermost part of the inclined channel.

3.2. Sample

The samples used in this study were obtained from the Panzhihua area of Sichuan, China. The raw material was tailings from ilmenite flotation, which, after impurity removal, grinding, classification, and desliming, became the experimental samples.

The size distribution of the feed material for test work was measured by a Malvern Mastersizer 2000 laser diffractometer (Comprehensive Laboratory, Resource and Biological Experiment Center, School of Resources Processing and Bioengineering, Central South University, Changsha, China). Particle size is reported as equivalent spherical diameter. Figure 5 illustrates that the sample's particle size distribution ranges within −80 μm, with roughly 50% of the particles measuring finer than 13.8 μm in size. The samples were screened using 38 and 18 μm sieves, and the mass percentages and TiO_2 grade are shown in Table 1. It can be seen from the table that the majority of particles were less than 18 μm. The material in the sample with a particle size less than 18 μm has the highest proportion, accounting for 56.96%, with a TiO_2 grade of 11.62% and a TiO_2 distribution rate of 75.46%. The majority of the Ti elements are present in the −18 μm fraction, which poses a significant challenge for gravity separation technology.

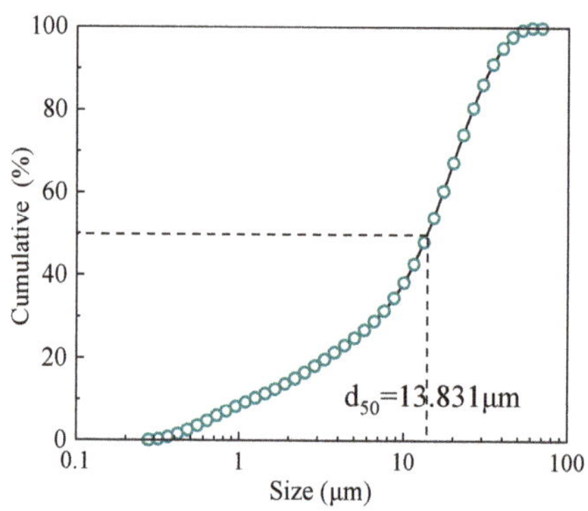

Figure 5. Size analysis results.

Table 1. Sieve analysis result of feed.

Size (μm)	Mass Fraction (%)	TiO_2 Grade (%)	TiO_2 Distribution (%)
+38	11.58	2.43	3.21
−38 + 18	31.46	5.94	21.32
−18	56.96	11.62	75.46
Total	100	8.77	100

The result of X-ray diffraction (XRD) analysis is shown in Figure 6, XRD analysis of the samples revealed a complex mineral composition, consisting of ilmenite as the target mineral, and augite, chlorite, feldspar, and amphibole as the primary gangue minerals.

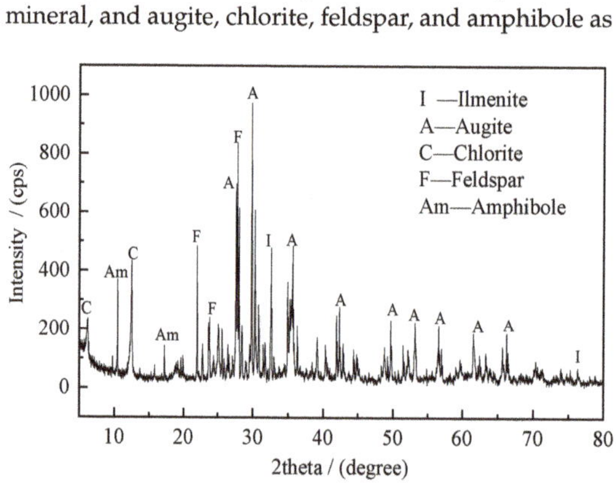

Figure 6. X-ray diffraction pattern of mineral sample.

3.3. Method

Before conducting the experiment, it is essential to prepare a precise mass concentration of the slurry in a tank. A stirrer is employed to prevent any potential issues related to settling and layering. Once the fluidization water is activated, and the throughput and underflow rate are adjusted, the feed pump can be launched to supply the slurry to the equipment. The operating procedure of the RC is depicted in Figure 7.

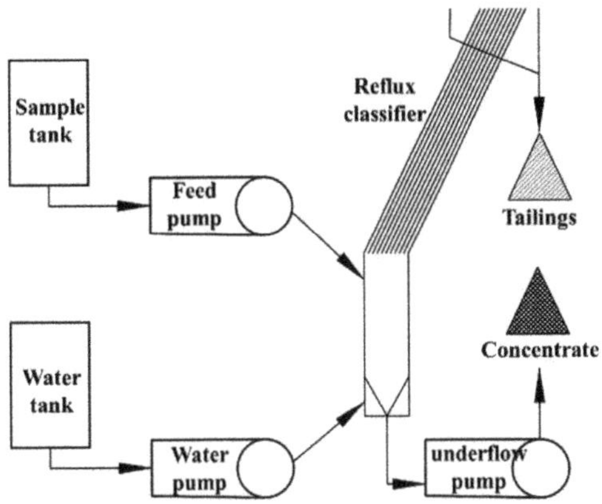

Figure 7. Experimental operating procedure.

One hour after the feed pump starts, the equipment reaches a completely stable state. At this juncture, the relationship between the mass and grade of the product and the feed is in equilibrium. During stable separation conditions of the RC, samples are collected from the fluidized bed, tailings, and concentrate. The particles of samples are weighed and subjected to chemical analysis to calculate the separation efficiency.

The control variates method is utilized to assess the performance of the RC in the processing of ultrafine ilmenite. The variables under investigation include the underflow rate, fluidization water velocity, and throughput. These three variables are regulated by the underflow pump, fluidization water pump, and feed pump, respectively.

Three consecutive experiments are conducted under the same experimental conditions, with each sampling interval lasting 60 min. The results are then summarized, and the standard deviation and mean values are calculated. The equation for calculating separation efficiency and recovery is given below:

$$R = \frac{C \cdot \beta}{\alpha}, \quad (7)$$

$$\eta = \frac{(\alpha - \varepsilon)(\beta - \alpha) \times 100}{\alpha(\beta - \varepsilon)(100 - \alpha)} \times 100\%, \quad (8)$$

where R is the recovery of the concentrated product, η is the separation efficiency [27], C is the yield of concentrate, and α, β, and ε are (in %) the TiO_2 grade of feed, concentrate, and tailings.

4. Results

4.1. Effects of Underflow Rate

The specific information of the experimental results is listed in Appendix A Tables A1–A3, which detail the mean results of multiple experiments. The experimental results for the underflow rate are displayed in Figure 8. As shown in Figure 8, the increase in the underflow rate will result in a synchronous decrease in the concentrate grade and tailings grade. More material enters the concentrate product, including ilmenite and gangue, which leads to a decrease in the grade and an increase in the recovery. As shown in Figure 8c, an increase in the underflow will lead to a reduction in bed density. According to Equations (1) and (2), the fact that the fluidization water has not changed but the density of the fluidized bed has decreased results in a smaller difference in particle settling velocities, which is the reason for the decrease in separation efficiency. Furthermore, rapid entrainment of heavy minerals into the underflow prevents effective enrichment within the bed. Consequently, this results in shorter separation times and a lower bed grade, which is the primary cause of the decrease in concentrate grade. An excessively high underflow yield could lead to insufficient separation based on density and result in incomplete separation of particles before being directed to the concentrate product, thereby decreasing the separation efficiency. The lowest underflow rate for this device is 17.00 mL/min, and a lower rate will lead to blockages and abnormal operation.

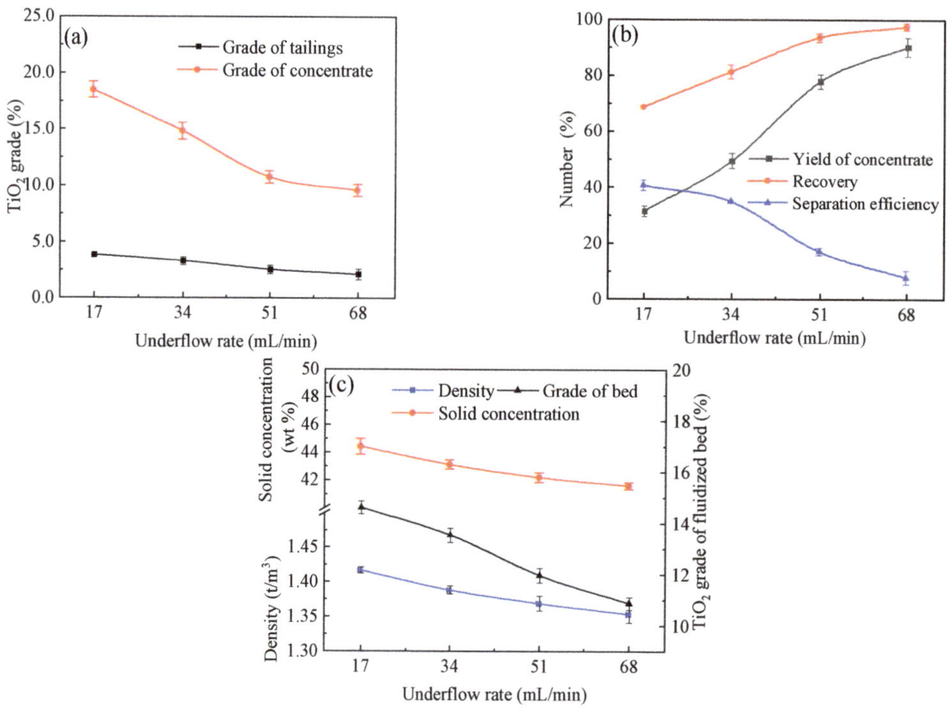

Figure 8. The effect of underflow rate on the performance of RC (**a**) TiO_2 grade of products; (**b**) recovery, yield of tailings, and separation efficiency; (**c**) concentration, density, and grade of fluidized bed (concentration of feed = 30%, throughput = 0.71 t/(m^2·h), fluidization water velocity = 2.67×10^{-3} m/s).

4.2. *Effects of Fluidization Water Velocity*

The fluidization water velocity mainly affects the drag force exerted on the particles and the solid concentration in the fluidized bed. An increase in fluidization water velocity will cause fine mineral particles to enter the overflow product more easily, while coarser gangue minerals are more likely to be discarded. Therefore, the fluidization water velocity needs to be experimentally determined.

As can be seen in Figure 9a,b, consistent with the pattern observed in the calculations, an increase in fluidization water velocity results in a marked reduction in the recovery and an increase in the quantity of solid entering the overflow. Additionally, concentrate and tailings grades were both improved with the escalation of fluidization water velocity.

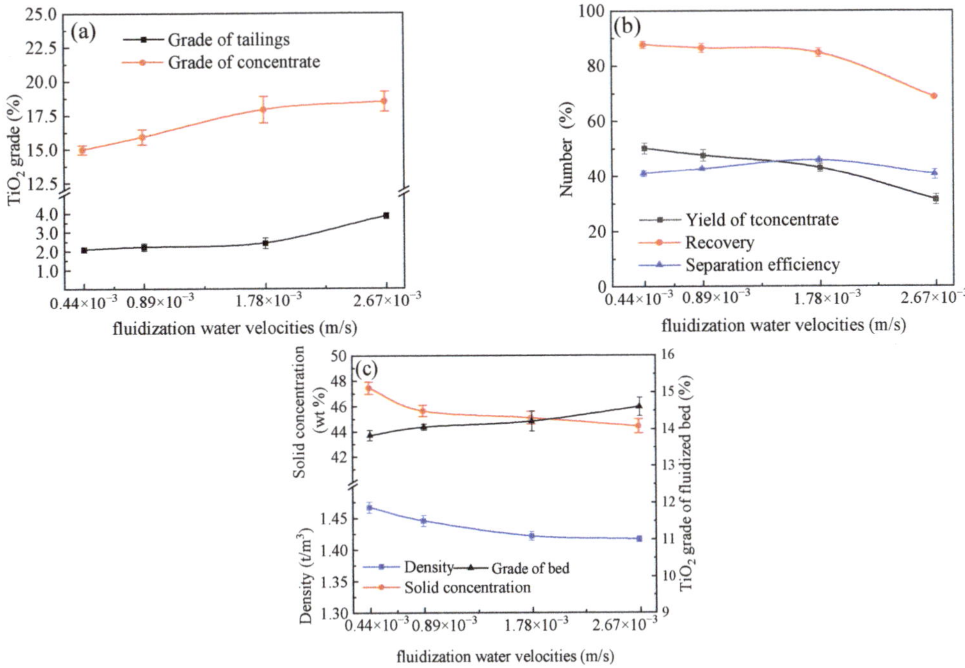

Figure 9. The effect of fluidization water velocity on the performance of RC (**a**) TiO_2 grade of products; (**b**) recovery, yield of tailings, and separation efficiency; (**c**) concentration, density, and grade of fluidized bed (solids concentration of feed = 30%, throughput = 0.71 t/(m^2·h), underflow flux = 17.00 mL/min).

As shown in Figure 9c, an increase in the fluidized water velocity results in the dilution of the fluidized bed, causing a simultaneous decrease in fluidized bed density and concentration. According to Equation (1), as the fluidization water velocity increases, lighter particles encounter more difficulty in settling into the fluidized bed, leading to a higher TiO_2 grade within the fluidized bed. At the same time, an increase in the fluidization water velocity also leads to an increase in the slurry velocity within the inclined channels. According to Equation (6), the *Lp* values of various particles increase, especially the fine, heavy particles, which are more likely to be lost in the overflow. This results in an increase in tailings yield and a decrease in recovery rate. However, this also results in the loss of finer heavy particles. Therefore, an increase in fluidization water velocity leads to an increase in concentrate grade but a decrease in recovery.

By computing the separation efficiency, it can be deduced that the highest efficiency emerges at a fluidization water velocity of 1.78×10^{-3} m/s.

4.3. Effects of Throughput

The throughput refers to the amount of solid material, measured in metric tonnes, that is processed per unit of horizontal vessel area per hour. This unit is usually expressed as square meters of horizontal cross-sectional area. As shown in Figure 10, an increase in throughput results in more particles being processed in the vessel, while also causing more heavy particles to enter the fluidized bed and the inclined channel over the same period. The overflow increases as the processing capacity increases. In addition, a greater mass of heavy minerals per unit area is deposited in the inclined channel. This results in a greater loss of heavy particles in the overflow, leading to a decrease in recovery. However, due to the enrichment of heavy particles in the fluidized bed, the concentration and grade of the fluidized bed are improved, thus resulting in an increase in the concentrate grade.

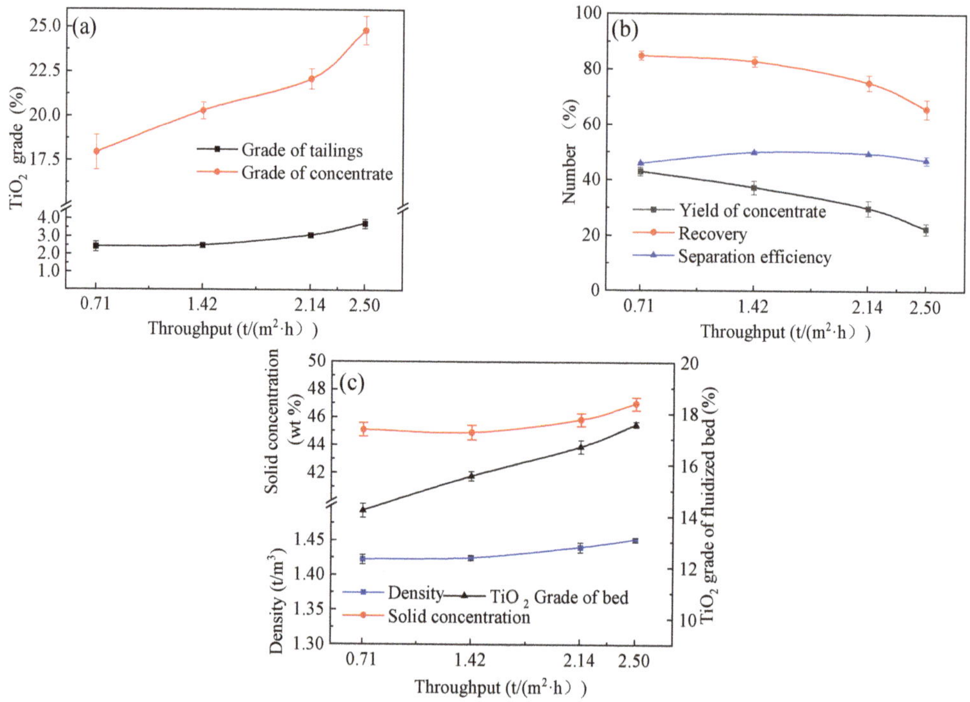

Figure 10. The effect of throughput on the performance of RC (**a**) TiO_2 grade of products; (**b**) recovery, yield of tailings, and separation efficiency; (**c**) concentration, density, and grade of fluidized bed (solids concentration of feed = 30%, fluidization water velocity = 1.78×10^{-3} m/s, underflow flux = 17.00 mL/min).

In Figure 10, an increase in throughput leads to a simultaneous increase in concentrate grade and tailings grade, while the recovery rate decreases significantly with an increase in tailings yield. The highest separation efficiency can be achieved when the throughput is 1.42 t/(m^2·h). At this point, the fluidized bed density is 1.425 t/m^3. Hence, in the processing of ultrafine ilmenite ore, the attainment of a notably high separation efficiency is observed at a fluidized bed density of approximately 1.425 t/m^3. This results in a higher separation efficiency of 50.0%, accompanied by an impressive concentrate recovery of 82.8% with a TiO_2 grade in the concentrate of 20.3%, while the tailings yield is 62.6%, with a TiO_2 grade in the tailings of 2.51%.

4.4. Comparison Between Magnetic Separator and RC

Magnetic separation is one of the primary methods for recovering and concentrating ilmenite. In order to draw comparisons, the feed sample of the RC was tested using an XCRQ—50 × 70 wet high-intensity Jones magnetic separator produced by Wuhan Hengle Mineral Engineering Equipment Co., Ltd. (Wuhan, China). The solid concentration of feed for the magnetic separation experiment was 5%, with unloading taking place every 5 s. The clamp was demagnetized during unloading, after which the particles captured by the magnetic medium were washed into the concentrate collection bucket.

Figure 11 shows the effect of the magnetic intensity on the separation performance of the Jones magnetic separator. The magnetic intensity described in the figure is the highest magnetic intensity on the magnetic medium plate teeth in the Jones magnetic separator, the spacing between the magnetic medium teeth is 1.5 mm, and the tooth angle is 45°. Figure 11 indicates that as the magnetic intensity increases, the recovery gradually improves while the tailings yield steadily decreases. At a magnetic intensity of 1.0 T, the concentrate can attain a maximum TiO_2 grade of 15.4%, yet the recovery rate is relatively low, at only 61.2%. Despite achieving a higher recovery rate by increasing the magnetic intensity to its maximum of 2.2 T, the concentrate grade is lower, resulting in ineffective enrichment.

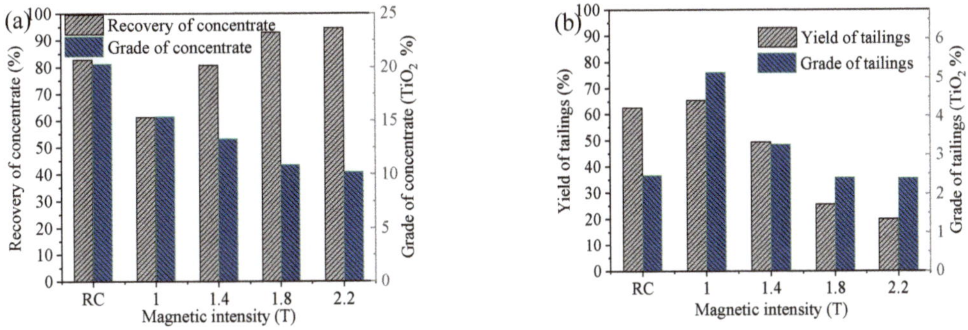

Figure 11. The effect of magnetic intensity (**a**) concentrate; (**b**) tailings.

Figure 11 also includes the gravity separation index, which was derived from the optimal performance in Section 4.3. As illustrated, at a magnetic intensity of 1.4 T, recovery similar to those obtained through gravity separation, approximately 80%, can be achieved. However, the concentrate TiO_2 grade obtained through gravity separation is higher, about 1.5 times that of magnetic separation. Additionally, the tailings grade of RC is relatively lower compared to the condition of a magnetic intensity of 1.4 T, with a higher yield. This indicates that the effect of RC is superior when processing the ilmenite samples.

XRD analysis was performed on the concentrate products obtained through gravity separation and magnetic separation, and the results are shown in Figure 12. As depicted, the peak heights of augite and chlorite increased significantly as the magnetic intensity increased. This indicates that the augite and chlorite in the feed sample possess certain magnetic properties. With an increase in magnetic intensity, gangue minerals will be captured onto the magnetic medium together with the ilmenite, making it difficult to achieve the separation of gangue minerals from ilmenite.

In the XRD curve of the concentrate products of RC, the peak heights of gangue minerals, particularly chlorite and augite, are relatively low. This is due to the significant difference in density between ilmenite (4450 kg/m^3) and gangue minerals such as augite (3230 kg/m^3), chlorite (3210 kg/m^3), and feldspar (2600 kg/m^3) in the feed sample. Therefore, the density difference was utilized as the basis for separation in gravity separation. Consequently, the density difference between gangue minerals and ilmenite resulted in a more favorable separation effect during the reprocessing.

When the highest separation efficiency is achieved through gravity separation, the recovery rate is similar to that achieved by magnetic separation with a magnetic field intensity of 1.4 T. During magnetic separation, a higher recovery rate can be achieved with a magnetic field intensity of 2.2 T. Figure 13 presents information on various particle sizes of products at magnetic intensities of 1.4 T and 2.2 T and the RC highest separation efficiency.

As seen in Figure 13b,c, increasing the magnetic field intensity significantly improved the concentrate yield and recovery of all size particles. The stronger magnetic force enhanced the recovery of coarse particles that are more susceptible to loss. However, due to the interference of magnetic gangue minerals, as seen in Figure 13a, a large number of particles were adsorbed onto the medium and thus entered the concentrate products, resulting in poor enrichment of ilmenite.

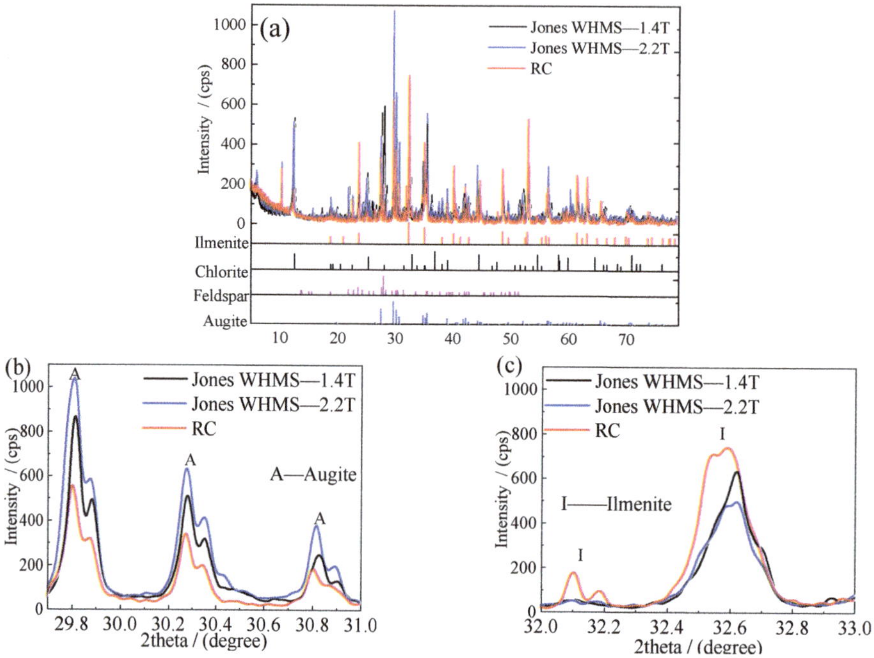

Figure 12. The X-ray diffraction analysis of the concentrate prepared by RC and magnetic separator (**a**) total diagram; (**b**) augite; (**c**) ilmenite.

By raising the magnetic intensity from 1.4 T to 2.2 T, the yield of the +38 μm coarse particles in the concentrate increased from 39.9% to 89.2%, while the grade decreased from 5.68% to 3.38%. The grade of the +38 μm particles in the feed was 2.43%. Hence, excessively high magnetic field intensity will result in a reduction in the selectivity of +38 μm particles.

As seen in Figure 13, increasing the magnetic intensity will similarly result in an elevation of the recovery rate of ultrafine ilmenite of −18 μm and can effectively reduce the content of TiO_2 in tailings products, as was reported in the introduction. However, it also leads to a selective reduction in which the grade of the concentrate is not significantly enhanced relative to the feed. Therefore, when employing magnetic separation to process this sample, lower magnetic field strength will drastically increase the tailings yield, especially for fine particles of −18 μm, which will cause a significant reduction in recovery. On the other hand, increasing the magnetic field strength will result in a reduction in separation efficiency, making it difficult to achieve effective concentration of concentrate products.

In contrast, as shown in Figure 13a, RC achieved a relatively high concentrate grade for all particle sizes and demonstrated a high enrichment ratio for the ultrafine −18 μm particles. As for the −18 μm material, the tailings grade (Figure 13d) generated by gravity separation was the same as the results obtained by magnetic separation at the highest intensity, but with a higher yield. Therefore, it can be seen that gravity separation has a good effect on the separation of ultrafine ilmenite minerals.

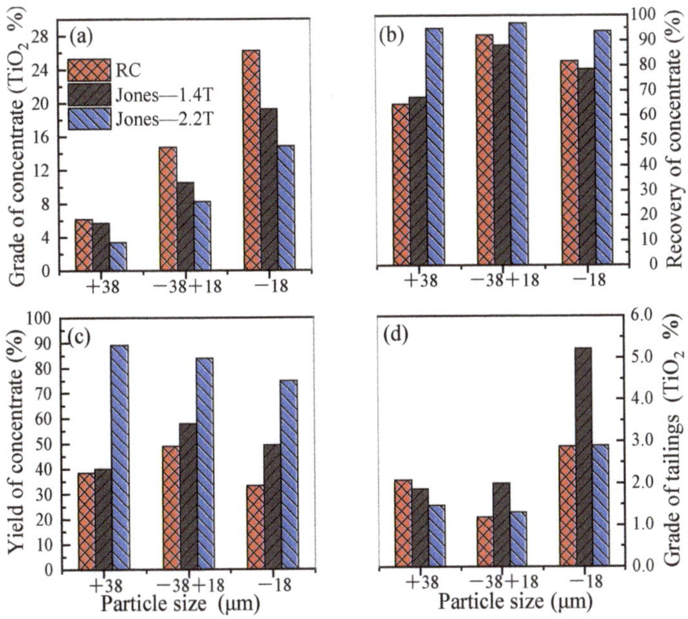

Figure 13. Separation performance of RC and magnetic separator at various particle sizes. (**a**) Grade of concentrate; (**b**) recovery of concentrate; (**c**) yield of concentrate; (**d**) grade of tailings.

5. Conclusions

The efficient and cost-effective preconcentration of fine-grained ilmenite presents a challenge, as traditional beneficiation techniques are often prohibitively expensive or difficult to implement effectively. However, experimentation has demonstrated that the reflux classifier (RC) is highly effective for the preconcentration of low-grade ultrafine ilmenite.

1. The results indicate that using the reflux classifier for the preconcentration of ultrafine ilmenite ore, with an underflow rate of 17 mL/min, a fluidization water velocity of 1.78×10^{-3} m/s, and a throughput of 1.42 t/(m^2·h), yields a maximum achievable separation efficiency of 50.0%. Under these conditions, the concentrate grade of TiO_2 is 20.3%, with a recovery rate of 82.8%. The tailings yield is 62.6%, with a tailings TiO_2 grade of 2.44%.
2. A comparative analysis of the beneficiation results between magnetic separation and the reflux classifier reveals that certain gangue minerals in the ore, such as augite and chlorite, exhibit some degree of magnetism and can be captured by the magnetic medium due to increased magnetic intensity. This negatively impacts the quality of the concentrate product. However, the magnetic properties of the gangue do not affect the efficiency of the gravity process. For this specific material, magnetic separation was ineffective, while the reflux classifier achieved favorable results.
3. It is noteworthy that the ore sample used in the experiment was exceptionally fine-grained, with 50% of the material having particle sizes finer than 13.8 μm. Despite this, the reflux classifier demonstrated commendable performance, highlighting the significant potential of this technology for the cost-effective and efficient preconcentration of ultrafine ilmenite ore.

Author Contributions: Conceptualization, D.L. and Z.L.; methodology, D.L. and Z.L.; validation, Z.L. and Y.Z.; formal analysis, Z.L.; data curation, Z.L.; writing—original draft preparation, Z.L.;

writing—review and editing, Z.L., D.L., Y.W., Z.S., B.L., X.Z., K.C. and X.H.; funding acquisition, D.L. and Y.W. All authors have read and agreed to the published version of the manuscript.

Funding: The research was supported by the National Key Research and Development Program of China (No. 2021YFC2903202) and the National Natural Science Foundation of China (52174270, 52174267, 51974366, and 51674290).

Data Availability Statement: Data are contained within the article.

Conflicts of Interest: Zhenhua Su, Bing Liu and Xuqun Zhong are employees of Guangxi Zhongiin Lingnan Mining Co., Ltd. The paper reflects the views of the scientists and not the company.

Appendix A

Table A1. Summary of conditions used in the experimental program.

Run	Underflow Rate (mL/min)	Fluidization Water Velocity (m/s)	Throughput (t/(m^2·h))	Concentration of Feed (%)
1	17.00	2.67×10^{-3}	0.71	30.00
2	34.00	2.67×10^{-3}	0.71	30.00
3	51.00	2.67×10^{-3}	0.71	30.00
4	68.00	2.67×10^{-3}	0.71	30.00
5	17.00	4.44×10^{-4}	0.71	30.00
6	17.00	8.89×10^{-4}	0.71	30.00
7	17.00	1.78×10^{-3}	0.71	30.00
8	17.00	1.78×10^{-3}	1.42	30.00
9	17.00	1.78×10^{-3}	2.14	30.00
10	17.00	1.78×10^{-3}	2.50	30.00

Table A2. Summary of product results of the experimental program.

Run	Grade of Tailings (%)	Grade of Concentrate (%)	Yield of Concentrate (%)	Recovery (%)	Separation Efficiency (%)
1	3.85	18.51	31.47	68.78	40.76
2	3.30	14.83	49.50	81.47	35.12
3	2.54	10.78	77.98	93.71	17.28
4	2.13	9.61	90.34	97.59	7.96
5	2.10	15.03	50.23	87.80	41.10
6	2.23	15.96	47.61	86.61	42.75
7	2.44	17.94	43.02	84.75	45.90
8	2.51	20.29	37.39	82.80	49.98
9	3.11	22.11	30.06	75.26	49.57
10	3.77	24.84	22.72	65.89	47.21

Table A3. Summary of fluidized-bed results of the experimental program.

Run	Density (t/m^3)	Solid Concentration (wt %)	TiO_2 Grade of Bed (%)
1	1.42	44.43	14.60
2	1.39	43.14	13.53
3	1.37	42.23	11.96
4	1.35	41.62	10.87
5	1.47	47.47	13.84
6	1.45	45.65	14.06
7	1.42	45.08	14.21
8	1.42	44.89	15.55
9	1.44	45.83	16.70
10	1.45	47.00	17.57

References

1. Wu, F.; Li, X.; Wang, Z.; Wu, L.; Guo, H.J.; Xiong, X.H.; Zhang, X.P.; Wang, X.J. Hydrogen peroxide leaching of hydrolyzed titania residue prepared from mechanically activated Panzhihua ilmenite leached by hydrochloric acid. *Int. J. Miner. Process.* **2011**, *98*, 106–112. [CrossRef]
2. Liu, S.-s.; Guo, Y.-f.; Qiu, G.-z.; Jiang, T.; Chen, F. Solid-state reduction kinetics and mechanism of pre-oxidized vanadium-titanium magnetite concentrate. *Trans. Nonferrous Met. Soc. China* **2014**, *24*, 3372–3377. [CrossRef]
3. Yang, Y.; Xu, L.; Deng, J.; Liu, S. Influence of particle size on flotation separation of ilmenite, olivine, and pyroxene. *Politech. Wroc. Oficyna Wydawnicza* **2021**, *57*, 106–117. [CrossRef]
4. Lin, L.I. Experimental Research on Comprehensive Recovery of Iron and Titanium from the Low Grade Vanadium Titanium Magnetite Ore in Panzhihua. *Conserv. Util. Miner. Resour.* **2015**, *2*, 27–32. [CrossRef]
5. Liu, W.; Zhang, J.; Wang, W.; Deng, J.; Chen, B.; Yan, W.; Xiong, S.; Huang, Y.; Liu, J. Flotation behaviors of ilmenite, titanaugite, and forsterite using sodium oleate as the collector. *Miner. Eng.* **2015**, *72*, 1–9. [CrossRef]
6. Zhang, H.; Zeng, J.; Xie, H.; Guan, C.; Chen, L. Enhanced separation for ilmenite tailings with a novel HGMS-flotation process. *Sep. Sci. Technol.* **2020**, *55*, 752–760. [CrossRef]
7. Shen, S.P.; Yuan, Z.T.; Liu, J.T.; Meng, Q.Y.; Hao, H.Q. Preconcentration of ultrafine ilmenite ore using a superconducting magnetic separator. *Powder Technol.* **2020**, *360*, 1–9. [CrossRef]
8. Xu, C.; Zhang, Y.; Liu, T.; Huang, J. Characterization and Pre-Concentration of Low-Grade Vanadium-Titanium Magnetite Ore. *Minerals* **2017**, *7*, 137. [CrossRef]
9. Laxmi, T.; Satya Sai, S.; Rao, D.S.; Rao, R.B. Beneficiation studies on recovery and in-depth characterization of ilmenite from red sediments of badlands topography of Ganjam District, Odisha, India. *Int. J. Min. Sci. Technol.* **2013**, *23*, 725–731. [CrossRef]
10. Chen, L.; Yang, R.; Zeng, J.; Shao, Y.; Xiao, Q.; Guo, S. A wet belt permanent high gradient magnetic separator for purification of non-metallic ores. *Int. J. Miner. Process.* **2016**, *153*, 66–70. [CrossRef]
11. Svoboda, J. The effect of magnetic field strenght on the efficiency of magnetic separation. *Miner. Eng.* **1994**, *7*, 747–757. [CrossRef]
12. Ge, W.; Encinas, A.; Araujo, E.; Song, S. Magnetic matrices used in high gradient magnetic separation (HGMS): A review. *Results Phys.* **2017**, *7*, 4278–4286. [CrossRef]
13. Li, L.; Shen, S.; Yuan, Z.; Wang, H. Loss mechanism of fine-grainedilmenite in magnetic separation. *China Min. Mag.* **2018**, *27*, 138–144. (In Chinese) [CrossRef]
14. Fan, G.; Zhang, C.; Wang, T.; Deng, J.; Cao, Y.; Chang, L.; Zhou, G.; Wu, Y.; Li, P. New insight into surface adsorption thermodynamic, kinetic properties and adsorption mechanisms of sodium oleate on ilmenite and titanaugite. *Adv. Powder Technol.* **2020**, *31*, 3628–3639. [CrossRef]
15. Fan, X.; Rowson, N.A. The effect of $Pb(NO_3)_2$ on ilmenite flotation. *Miner. Eng.* **2000**, *13*, S0892–S6875. [CrossRef]
16. Miettinen, T.; Ralston, J.; Fornasiero, D. The limits of fine particle flotation. *Miner. Eng.* **2010**, *23*, 420–437. [CrossRef]
17. Liu, G.Z.; Dai, S.J.; Bai, L.M.; Ma, Y.X.; Zhang, Y. Experimental Studies on a Mineral Containing Titanium in Baoding Area. *Adv. Mater. Res.* **2013**, *826*, 34–37. [CrossRef]
18. Galvin, K.P.; Zhou, J.; Walton, K. Application of closely spaced inclined channels in gravity separation of fine particles. *Miner. Eng.* **2010**, *23*, 326–338. [CrossRef]
19. Galvin, K.P.; Doroodchi, E.; Callen, A.M.; Lambert, N.; Pratten, S.J. Pilot plant trial of the reflux classifier. *Miner. Eng.* **2002**, *15*, 19–25. [CrossRef]
20. Galvin, K.P.; Callen, A.; Zhou, J.; Doroodchi, E. Performance of the reflux classifier for gravity separation at full scale. *Miner. Eng.* **2005**, *18*, 19–24. [CrossRef]
21. Nguyentranlam, G.; Galvin, K.P. Applications of the Reflux Classifier in solid-liquid operations. *Int. J. Miner. Process.* **2004**, *73*, 83–89. [CrossRef]
22. Galvin, K.P.; Callen, A.M.; Spear, S. Gravity separation of coarse particles using the Reflux Classifier. *Miner. Eng.* **2010**, *23*, 339–349. [CrossRef]
23. Amariei, D.; Michaud, D.; Paquet, G.; Lindsay, M. The use of a Reflux Classifier for iron ores: Assessment of fine particles recovery at pilot scale. *Miner. Eng.* **2014**, *62*, 66–73. [CrossRef]
24. Hunter, D.M.; Iveson, S.M.; Galvin, K.P. A batch elutriation technique for the density-based fractionation of iron ore. In Proceedings of the Iron Ore 2017, Perth, Australia, 24–26 July 2017.
25. Rodrigues, A.F.d.V.; Delboni Junior, H.; Rodrigues, O.M.S.; Zhou, J.; Galvin, K.P. Gravity separation of fine itabirite iron ore using the Reflux Classifier-Part I-Investigation of continuous steady state separations across a wide range of parameters. *Miner. Eng.* **2023**, *201*, 108187. [CrossRef]
26. Richardson, J.F.; Zaki, W.N. Sedimentation and fluidisation: Part I. *Chem. Eng. Res. Des.* **1997**, *75*, S0263–S8762. [CrossRef]
27. Xie, G. *Mineral Processing*; University of Mining and Technology Press: Xuzhou, China, 2016.
28. Galvin, K.P.; Liu, H. Role of inertial lift in elutriating particles according to their density. *Chem. Eng. Sci.* **2011**, *66*, 3687–3691. [CrossRef]

29. King, M.R.; Leighton, J.D.T. Measurement of the inertial lift on a moving sphere in contact with a plane wall in a shear flow. *Phy Fluids* **1997**, *9*, 1248–1255. [CrossRef]
30. Vance, W.H.; Moulton, R.W. A study of slip ratios for the flow of steam-water mixtures at high void fractions. *AIChE J.* **1965**, *1* 1114–1124. [CrossRef]
31. Li, Y.; Li, Y.; Xia, W.; He, C.Y.; Zhu, R.T. A novel particulate sedimentation model in inclined channels of liquid-solid fluidize bed. *Powder Technol.* **2017**, *305*, 764–770. [CrossRef]

MDPI AG
Grosspeteranlage 5
4052 Basel
Switzerland
Tel.: +41 61 683 77 34

Minerals Editorial Office
E-mail: minerals@mdpi.com
www.mdpi.com/journal/minerals

www.ingramcontent.com/pod-product-compliance
Lightning Source LLC
Chambersburg PA
CBHW061614210926
51850CB00009B/2325